OBJECTIVE SOIL PHYSICS

(*for JRF, SRF, ARS, NET, SLET, Civil Services & Other Competitive Examinations*)

OBJECTIVE SOIL PHYSICS

(for JRF, SRF, ARS, NET, SLET, Civil Services & Other Competitive Examinations)

Shashi Bhushan Kumar
Assistant Professor-cum-Junior Scientist, BAU, Ranchi, Jharkhand

Brajendra
Senior Scientist, Soil Science, Directorate of Rice Research, Hyderabad

A.K. Vishwakarma
Scientist, Agronomy, ICAR Research Complex for NEH Region, Umiam, Meghalaya

T. Ramesh
Scientist, Soil Science, ICAR Research Complex for NEH Region, Umiam, Meghalaya

Ashok Kumar
Programme Coordinator, Krishi Vigyan Kendra, Baliapur, Dhanbad, Jharkhand

Daya Publishing House®
A Division of
Astral International Pvt. Ltd.
New Delhi – 110 002

ISBN: 978-81-7035-662-2 (Paperback)

Publisher's Note:

Published by : **Daya Publishing House®**
A Division of
Astral International Pvt. Ltd.
– ISO 9001:2015 Certified Company –
4736/23, Ansari Road, Darya Ganj
New Delhi-110 002
Ph. 011-43549197, 23278134
E-mail: info@astralint.com
Website: www.astralint.com

Acknowledgements

Authors sincerely thank the help rendered by several persons in the present dissertation. We are indebted to the help rendered by Dr. N.V.K. Chakravorty Ex. Head Agril. Physics Division, IARI, New Delhi, Dr. Usha Kiran Chopra, Professor Agril. Physics Division, IARI, New Delhi, Dr. Debashis Chakravorty and Dr. R. N. Sahoo, Scientists Agril. Physics Division, IARI, New Delhi for their guidance in several ways. The past and present students of Agricultural Physics division are also acknowledged for their suggestions. Our thanks are also due to the course instructors, course leaders and course coordinators of winter school on Remote sensing applications in Agriculture in the Division of Agricultural Physics, IARI, New Delhi for their beautiful deliberations of the subject. Last but not least, beautiful type setting and printing works done by the publisher Daya Publishing House is highly acknowledged. Authors also acknowledge one and all for the help rendered.

Authors

Contents

Chapter 1
Units and Dimensions

Q. 1 One femto is equal to:

(*a*) 10^{-15} (*b*) 10^{-13}

(*c*) 10^{-12} (*d*) 10^{-10}

Q. 2 One Tera is equal to:

(*a*) 10^{12} (*b*) 10^{11}

(*c*) 10^{5} (*d*) 10^{10}

Q. 3 Which of the following is correct for one liter volume:

(*a*) A liter is equal to the volume of a cube exactly 10 cm tall, 10 cm long, and 10 cm wide.

(*b*) Volume of a cube exactly 1 cm tall, 1 cm long, and 1 cm wide.

(*c*) A liter is equal to the volume of a cube exactly 100 cm tall, 100 cm long, and 100 cm wide.

(*d*) All of the above

Q. 4 In which year the International System of Units was proposed as a replacement for the Metric System:

(*a*) 1960 (*b*) 1970

(*c*) 1980 (*d*) 2000

Q. 5 Candela is the unit of:

(*a*) Luminous intensity (*b*) Light year

(*c*) Energy (*d*) Mass

Q. 6 How many base units are there in SI units:

(*a*) 3 (*b*) 4

(*c*) 5 (*d*) 7

Q. 7 Electric potential can be expressed as:

(*a*) V (*b*) J/C

(*c*) Both (*d*) None

Q. 8 Energy can be expressed as:

(*a*) J (*b*) kg-m^2/s^2

(*c*) Both (*d*) None

Q. 9 Force can be expressed as:

(*a*) N (*b*) kg-m/s^2

(*c*) Both (*d*) None

Q. 10 Pressure can also be expressed as:

(*a*) Pa (*b*) (N/m^2)

(*c*) Both (*d*) None

Q. 11 Torr is a SI unit of pressure (T/F)

(*a*) True (*b*) False

Q. 12 mmHg is equal to:

(*a*) 133.32 Pa (*b*) atm

(*c*) 101.325 kPa (*d*) All of the above

Q. 13 SI Derived Units of Frequency:

(*a*) 1/s (*b*) m kg/s^2

(*c*) $kg/m\,s^2$ (*d*) $m^2\,kg/s^2$

Q. 14 SI Derived Units of force:

(*a*) 1/s (*b*) m kg/s^2

(*c*) $kg/m\,s^2$ (*d*) $m^2\,kg/s^2$

Q. 15 SI Derived Units of Pressure:

(*a*) 1/s (*b*) m kg/s^2

(*c*) $kg/m\,s^2$ (*d*) $m^2\,kg/s^2$

Q. 16 SI Derived Units of Stress:

(*a*) 1/s (*b*) m kg/s^2

(*c*) kg/m s^2 (*d*) m^2 kg/s^2

Q. 17 SI Derived Units of Power, radiant flux:

(*a*) m^2 kg/s^3 (*b*) s A

(*c*) m^2 kg/s^3 A (*d*) s^4 A^2/m^2 kg

Q. 18 SI Derived Units of Quantity of electricity, electric charge:

(*a*) m^2 kg/s^3 (*b*) s A

(*c*) kg/s^3 A (*d*) s^4 A^2/m^2 kg

Q. 19 SI Derived Units of Electric potential:

(*a*) m^2 kg/s^3 (*b*) s A

(*c*) m^2 kg/s^3 A (*d*) s^4 A^2/m^2 kg

Q. 20 SI Derived Units of Capacitance:

(*a*) m^2 kg/s^3 (*b*) s A

(*c*) m^2 kg/s^3 A (*d*) s^4 A^2/m^2 kg

Q. 21 SI Derived Units of Electric resistance:

(*a*) m^2 kg/s^3 A^2 (*b*) s^3 A^2/m^2 kg

(*c*) m^2 kg/s^2 A (*d*) kg/s^2 A

Q. 22 SI Derived Units of Conductance:

(*a*) m^2 kg/s^3 A^2 (*b*) s^3 A^2/m^2 kg

(*c*) m^2 kg/s^2 A (*d*) kg/s^2 A

Q. 23 SI Derived Units of Magnetic flux:

(*a*) m^2 kg/s^3 A^2 (*b*) s^3 A^2/m^2 kg

(*c*) m^2 kg/s^2 A (*d*) kg/s^2 A

Q. 24 SI Derived Units of Magnetic flux density, magnetic induction:

(*a*) m^2 kg/s^3 A^2 (*b*) s^3 A^2/m^2 kg

(*c*) m^2 kg/s^2 A (*d*) kg/s^2 A

Q. 25 SI Derived Units of Inductance:

(*a*) $m^2 kg/s^2 A^2$ (*b*) cd sr

(*c*) cd sr/m^2 (*d*) Bq

Q. 26 SI Derived Units of Luminous flux:

(*a*) $m^2 kg/s^2 A^2$ (*b*) cd sr

(*c*) cd sr/m^2 (*d*) Bq

Q. 27 SI Derived Units of Illuminance:

(*a*) $m^2 kg/s^2 A^2$ (*b*) cd sr

(*c*) cd sr/m^2 (*d*) Bq

Q. 28 SI Derived Units of Activity (ionizing radiations):

(*a*) $m^2 kg/s^2 A^2$ (*b*) cd sr

(*c*) cd sr/m^2 (*d*) Bq

Q. 29 SI Derived Units of Dynamic viscosity:

(*a*) Pascal second: Pa s = kg/m s

(*b*) Metre newton: N m = $m^2 kg/s^2$

(*c*) Newton per metre: N/m = kg/s^2

(*d*) Watt per square metre: W/m^2 = kg/s^3

Q. 30 SI Derived Units of Moment of force:

(*a*) Pascal second: Pa s = kg/m s

(*b*) Metre newton: N m = $m^2 kg/s^2$

(*c*) Newton per metre: N/m = kg/s^2

(*d*) Watt per square metre: W/m^2 = kg/s^3

Q. 31 SI Derived Units of Surface tension:

(*a*) Pascal second: Pa s = kg/m s

(*b*) Metre newton: N m = $m^2 kg/s^2$

(*c*) Newton per metre: N/m = kg/s^2

(*d*) Watt per square metre: W/m^2 = kg/s^3

Q. 32 SI Derived Units of Heat flux density, irradiance:

(*a*) Pascal second: Pa s = kg/m s

(*b*) Metre newton: N m = $m^2 kg/s^2$

(*c*) Newton per metre: $N/m = kg/s^2$

(*d*) Watt per square metre: $W/m^2 = kg/s^3$

Q. 33 SI Derived Units of Heat capacity, entropy:

(*a*) Joule per kelvin: $J/K = m^2\,kg/s^2\,K$

(*b*) Joule per kilogram kelvin: $J/kg\,K = m^2/s^2\,K$

(*c*) Joule per kilogram: $J/kg = m^2/s^2$

(*d*) Watt per metre kelvin: $W/m\,K = m\,kg/s^3\,K$

Q. 34 SI Derived Units of Specific heat capacity, specific entropy:

(*a*) Joule per kelvin: $J/K = m^2\,kg/s^2\,K$

(*b*) Joule per kilogram kelvin: $J/kg\,K = m^2/s^2\,K$

(*c*) Joule per kilogram: $J/kg = m^2/s^2$

(*d*) Watt per metre kelvin: $W/m\,K = m\,kg/s^3\,K$

Q. 35 SI Derived Units of Specific energy:

(*a*) Joule per kelvin: $J/K = m^2\,kg/s^2\,K$

(*b*) Joule per kilogram kelvin: $J/kg\,K = m^2/s^2\,K$

(*c*) Joule per kilogram: $J/kg = m^2/s^2$

(*d*) Watt per metre kelvin: $W/m\,K = m\,kg/s^3\,K$

Q. 36 SI Derived Units of Thermal conductivity:

(*a*) Joule per kelvin: $J/K = m^2\,kg/s^2\,K$

(*b*) Joule per kilogram kelvin: $J/kg\,K = m^2/s^2\,K$

(*c*) Joule per kilogram: $J/kg = m^2/s^2$

(*d*) Watt per metre kelvin: $W/m\,K = m\,kg/s^3\,K$

Q. 37 SI Derived Units of Energy density:

(*a*) Joule per cubic metre: $J/m^3 = kg/m\,s^2$

(*b*) Volt per metre: $V/m = m\,kg/s^3\,A$

(*c*) Coulomb per cubic metre: $C/m^3 = s\,A/m^3$

(*d*) Coulomb per square metre: $C/m^2 = s\,A/m^2$

Q. 38 SI Derived Units of Electric field strength:

(*a*) Joule per cubic metre: $J/m^3 = kg/m\,s^2$

(*b*) Volt per metre: $V/m = m\,kg/s^3\,A$

(*c*) Coulomb per cubic metre: $C/m^3 = s\,A/m^3$

(*d*) Coulomb per square metre: $C/m^2 = s\,A/m^2$

Q. 39 SI Derived Units of Electric charge density:

(*a*) Joule per cubic metre: $J/m^3 = kg/m\,s^2$

(*b*) Volt per metre: $V/m = m\,kg/s^3\,A$

(*c*) Coulomb per cubic metre: $C/m^3 = s\,A/m^3$

(*d*) Coulomb per square metre: $C/m^2 = s\,A/m^2$

Q. 40 SI Derived Units of Electric displacement, electric flux density:

(*a*) Joule per cubic metre: $J/m^3 = kg/m\,s^2$

(*b*) Volt per metre: $V/m = m\,kg/s^3\,A$

(*c*) Coulomb per cubic metre: $C/m^3 = s\,A/m^3$

(*d*) Coulomb per square metre: $C/m^2 = s\,A/m^2$

Q. 41 SI Derived Units of Permittivity:

(*a*) Farad per metre: $F/m = s^4\,A^2/m^3\,kg$

(*b*) Henry per metre: $H/m = m\,kg/s^2\,A^2$

(*c*) Joule per mole: $J/mol = m^2\,kg/s^2\,mol$

(*d*) Joule per mole kelvin: $J/mol\,K = m^2\,kg/s^2\,K\,mol$

Q. 42 SI Derived Units of Permeability:

(*a*) Farad per metre: $F/m = s^4\,A^2/m^3\,kg$

(*b*) Henry per metre: $H/m = m\,kg/s^2\,A^2$

(*c*) Joule per mole: $J/mol = m^2\,kg/s^2\,mol$

(*d*) Joule per mole kelvin: $J/mol\,K = m^2\,kg/s^2\,K\,mol$

Q. 43 SI Derived Units of Molar energy:

(*a*) Farad per metre: $F/m = s^4\,A^2/m^3\,kg$

(*b*) Henry per metre: $H/m = m\,kg/s^2\,A^2$

(*c*) Joule per mole: $J/mol = m^2\,kg/s^2\,mol$

(*d*) Joule per mole kelvin: $J/mol\,K = m^2\,kg/s^2\,K\,mol$

Q. 44 SI Derived Units of Molar entropy, molar heat capacity:

(*a*) Farad per metre: $F/m = s^4 A^2/m^3 kg$

(*b*) Henry per metre: $H/m = m kg/s^2 A^2$

(*c*) Joule per mole: $J/mol = m^2 kg/s^2 mol$

(*d*) Joule per mole kelvin: $J/mol K = m^2 kg/s^2 K mol$

Q. 45 SI Derived Units of poise:

(*a*) 1 P = 1 dyn s/cm^2 = 0.1 Pa s

(*b*) 1 St = 1 cm^2/s = 10^{-4} m^2/s

(*c*) 1 G = 10^{-4} T

(*d*) 1 Oe = (1000/(4 pi)) A/m

Q. 46 SI Derived Units of stokes:

(*a*) 1 P = 1 dyn s/cm^2 = 0.1 Pa s

(*b*) 1 St = 1 cm^2/s = 10^{-4} m^2/s

(*c*) 1 G = 10^{-4} T

(*d*) 1 Oe = (1000/(4 pi)) A/m

Q. 47 SI Derived Units of gauss:

(*a*) 1 P = 1 dyn s/cm^2 = 0.1 Pa s

(*b*) 1 St = 1 cm^2/s = 10^{-4} m^2/s

(*c*) 1 G = 10^{-4} T

(*d*) 1 Oe = (1000/(4 pi)) A/m

Q. 48 SI Derived Units of oersted:

(*a*) 1 P = 1 dyn s/cm^2 = 0.1 Pa s

(*b*) 1 St = 1 cm^2/s = 10^{-4} m^2/s

(*c*) 1 G = 10^{-4} T

(*d*) 1 Oe = (1000/(4 pi)) A/m

Q. 49 SI Derived Units of Maxwell:

(*a*) 1 Mx = 10^{-8} Wb

(*b*) 1 sb = 1 cd/cm^2 = 10^4 cd/m^2

(*c*) 1 ph = 10^4 lx

(*d*) 1 dyn = 10^{-5} N

Q. 50 SI Derived Units of stilb:

(*a*) 1 Mx = 10^{-8} Wb

(*b*) 1 sb = 1 cd/cm^2 = 10^4 cd/m^2

(*c*) 1 ph = 10^4 lx

(*d*) 1 dyn = 10^{-5} N

Q. 51 SI Derived Units of phot:

(*a*) 1 Mx = 10^{-8} Wb

(*b*) 1 sb = 1 cd/cm^2 = 10^4 cd/m^2

(*c*) 1 ph = 10^4 lx

(*d*) 1 dyn = 10^{-5} N

Q. 52 SI Derived Units of dyne:

(*a*) 1 Mx = 10^{-8} Wb

(*b*) 1 sb = 1 cd/cm^2 = 10^4 cd/m^2

(*c*) 1 ph = 10^4 lx

(*d*) 1 dyn = 10^{-5} N

Q. 53 Numerical value of Planck constant *h*:

(*a*) $6.6260755 \times 10^{-34}$ J·s (*b*) $1.05457266 \times 10^{-34}$ J·s

(*c*) Both (*d*) None

Q. 54 Numerical value of Boltzmann constant k_B:

(*a*) 1.380658×10^{-23} J/K (*b*) 8.617385×10^{-5} eV/K

(*c*) Both (*d*) None

Q. 55 Numerical value of Elementary charge *e*:

(*a*) $1.60217733 \times 10^{-19}$ C

(*b*) 2.99792458×10^{8} m/s

(*c*) $9.1093897 \times 10^{-31}$ kg

(*d*) $1.6726231 \times 10^{-27}$ kg

Q. 56 Numerical value of Speed of light *c*:

(*a*) $1.60217733 \times 10^{-19}$ C

(*b*) 2.99792458×10^{8} m/s

(c) $9.1093897 \times 10^{-31}$ kg

(d) $1.6726231 \times 10^{-27}$ kg

Q. 57 Numerical value of Electron rest mass m_e:

(a) $1.60217733 \times 10^{-19}$ C

(b) 2.99792458×10^{8} m/s

(c) $9.1093897 \times 10^{-31}$ kg

(d) $1.6726231 \times 10^{-27}$ kg

Q. 58 Numerical value of Proton rest mass m_p:

(a) $1.60217733 \times 10^{-19}$ C

(b) 2.99792458×10^{8} m/s

(c) $9.1093897 \times 10^{-31}$ kg

(d) $1.6726231 \times 10^{-27}$ kg

Q. 59 Numerical value of Bohr magneton μ_B:

(a) $9.2740154 \times 10^{-24}$ J/T (b) $5.0507866 \times 10^{-27}$ J/T

(c) 2.002319304386 (d) 1.7608592×10^{11} 1/s·T

Q. 60 Numerical value of Nuclear magneton μ_N:

(a) $9.2740154 \times 10^{-24}$ J/T (b) $5.0507866 \times 10^{-27}$ J/T

(c) 2.002319304386 (d) 1.7608592×10^{11} 1/s·T

Q. 61 Numerical value of Free electron g factor g_e:

(a) $9.2740154 \times 10^{-24}$ J/T (b) $5.0507866 \times 10^{-27}$ J/T

(c) 2.002319304386 (d) 1.7608592×10^{11} 1/s·T

Q. 62 Numerical value of Free electron gyromagnetic ratio γ_e:

(a) $9.2740154 \times 10^{-24}$ J/T (b) $5.0507866 \times 10^{-27}$ J/T

(c) 2.002319304386 (d) 1.7608592×10^{11} 1/s·T

Q. 63 Numerical value of Electron magnetic moment μ_e:

(a) $-9.2847701 \times 10^{-24}$ J/T (b) 2.67515255×10^{8} 1/s·T

(c) 42.576375 MHz/T (d) $1.41060761 \times 10^{-26}$ J/T

Q. 64 Numerical value of Proton gyromagnetic ratio (H_2O) γ_p:

(a) $-9.2847701 \times 10^{-24}$ J/T (b) 2.67515255×10^{8} 1/s·T

(c) 42.576375 MHz/T (d) $1.41060761 \times 10^{-26}$ J/T

Q. 65 Numerical value of Proton magnetic moment μ_p:

(*a*) $-9.2847701 \times 10^{-24}$ J/T (*b*) 2.67515255×10^{8} 1/s·T

(*c*) 42.576375 MHz/T (*d*) $1.41060761 \times 10^{-26}$ J/T

Q. 66 Numerical value of Proton-electron ratios:

(*a*) $m_p/m_e = 1836.152701$

(*b*) $\mu_e/\mu_p = 658.2106881$

(*c*) $\gamma_e/\gamma_p = 658.2275841$ (protons in water)

(*d*) All of the above

Q. 67 Numerical value of Charge-to-mass ratio for the electron e/m_e:

(*a*) 1.75880×10^{11} C/kg (*b*) 1.66054×10^{-27} kg

(*c*) 5.29177×10^{-11} m (*d*) 2.81792×10^{-15} m

Q. 68 Numerical value of Atomic mass unit amu:

(*a*) 1.75880×10^{11} C/kg (*b*) 1.66054×10^{-27} kg

(*c*) 5.29177×10^{-11} m (*d*) 2.81792×10^{-15} m

Q. 69 Numerical value of Bohr radius a_0:

(*a*) 1.75880×10^{11} C/kg (*b*) 1.66054×10^{-27} kg

(*c*) 5.29177×10^{-11} m (*d*) 2.81792×10^{-15} m

Q. 70 Numerical value of Electron radius r_e:

(*a*) 1.75880×10^{11} C/kg (*b*) 1.66054×10^{-27} kg

(*c*) 5.29177×10^{-11} m (*d*) 2.81792×10^{-15} m

Q. 71 Numerical value of Gas constant *R*:

(*a*) 8.31451 m^2·kg/s^2·K·mol (*b*) 22.41383 m^3/kmol

(*c*) 9.64846×10^{4} C/mol (*d*) 5.585

Q. 72 Numerical value of Molar volume V_{mol}:

(*a*) 8.31451 m^2·kg/s^2·K·mol (*b*) 22.41383 m^3/kmol

(*c*) 9.64846×10^{4} C/mol (*d*) 5.585

Q. 73 Numerical value of Faraday constant *F*:

(*a*) 8.31451 m^2·kg/s^2·K·mol (*b*) 22.41383 m^3/kmol

(*c*) 9.64846×10^{4} C/mol (*d*) 5.585

Q. 74 Numerical value of Proton g factor (Landé factor) g_H:

(*a*) 8.31451 m²·kg/s²·K·mol (*b*) 22.41383 m³/kmol

(*c*) 9.64846 × 10⁴ C/mol (*d*) 5.585

Q. 75 Numerical value of Gravitational constant G:

(*a*) $(6.673 + -0.010) \times 10^{-11}$ m³/kg·s²

(*b*) $6.67390 \cdot 10^{-11}$ m³/kg·s² + –0.0014 %

(*c*) $(6.6873 + -0.0094) \times 10^{-11}$ m³/kg·s²

(*d*) All of the above

Q. 76 Numerical value of Acceleration due to gravity *g*:

(*a*) 9.80665 m/s² (*b*) $2.42631 \cdot 10^{-12}$ m

(*c*) 2.625501×10^{6} J/mol (*d*) None

Q. 77 Numerical value of Compton wavelength of the electron λ_c:

(*a*) 9.80665 m/s² (*b*) 2.42631×10^{-12} m

(*c*) 2.625501×10^{6} J/mol (*d*) None

Q. 78 Numerical value of Atomic energy unit Hartree:

(*a*) 9.80665 m/s² (*b*) 2.42631×10^{-12} m

(*c*) 2.625501×10^{6} J/mol (*d*) None

(a) [illegible] (b) [illegible]

(c) [illegible] (d) [illegible]

(a) 6.673 [illegible]

(b) [illegible]

(c) [illegible]

(d) All of these

(a) [illegible] (b) [illegible]

(c) [illegible] (d) None

(a) [illegible] (b) [illegible]

(c) [illegible] (d) None

(a) [illegible] (b) [illegible]

(c) [illegible] (d) None

ANSWERS

1	(*a*)	21	(*a*)	41	(*a*)	61	(*c*)
2	(*a*)	22	(*b*)	42	(*b*)	62	(*d*)
3	(*a*)	23	(*c*)	43	(*c*)	63	(*a*)
4	(*a*)	24	(*d*)	44	(*d*)	64	(*b*)
5	(*a*)	25	(*a*)	45	(*a*)	65	(*d*)
6	(*d*)	26	(*b*)	46	(*b*)	66	(*d*)
7	(*c*)	27	(*c*)	47	(*c*)	67	(*a*)
8	(*c*)	28	(*d*)	48	(*d*)	68	(*b*)
9	(*c*)	29	(*a*)	49	(*a*)	69	(*c*)
10	(*c*)	30	(*b*)	50	(*b*)	70	(*d*)
11	(*b*)	31	(*c*)	51	(*c*)	71	(*a*)
12	(*a*)	32	(*d*)	52	(*d*)	72	(*b*)
13	(*a*)	33	(*a*)	53	(*a*)	73	(*c*)
14	(*b*)	34	(*b*)	54	(*c*)	74	(*d*)
15	(*c*)	35	(*c*)	55	(*a*)	75	(*d*)
16	(*c*)	36	(*d*)	56	(*b*)	76	(*a*)
17	(*a*)	37	(*a*)	57	(*c*)	77	(*b*)
18	(*b*)	38	(*b*)	58	(*d*)	78	(*c*)
19	(*c*)	39	(*c*)	59	(*a*)		
20	(*d*)	40	(*d*)	60	(*b*)		

Chapter 2
Biophysics

Q. 1 Energy of Vander Waal's force............

(*a*) 1 Kcal/mole (*b*) 10 Kcal/mole

(*c*) 100 Kcal/mole (*d*) 1000 Kcal/mole

Q. 2 Energy of hydrogen bond in H-OH is ....

(*a*) 0.72 Kcal/mole (*b*) 7.2 Kcal/mole

(*c*) 72 Kcal/mole (*d*) 720 Kcal/mole

Q. 3 The living cell contains .... % of water.

(*a*) 10–20 (*b*) 30–40

(*c*) 30–40 (*d*) 70–90

Q. 4 Hydrogen bond is best explained in relation to the water molecule, which is a typical:

(*a*) Electrical dipole (*b*) Covalent

(*c*) Coordinate (*d*) None

Q. 5 Who classified all molecules into four groups with respect to their ability to participate in H bonding:

(*a*) Pimental (*b*) McClellan

(*c*) Both (*d*) None

Q. 6 The bond length of H-H is:

(*a*) 0.74 A° (*b*) 2.72 A°

(*c*) 3.23 A° (*d*) 2.78 A°

Q. 7 The average hydrogen bond length of O-H ....O is:

(*a*) 0.272 A° (*b*) 2.72 A°

(*c*) 27.2 A° (*d*) 272 A°

Q. 8 The average hydrogen bond length of C-H.......O is:

(*a*) 0.323 A° (*b*) 3.23 A°

(*c*) 32.3 A° (*d*) 323 A°

Q. 9 The average hydrogen bond length of N-H.....O is:

(*a*) 0.293 A° (*b*) 2.93 A°

(*c*) 29.3 A° (*d*) 293 A°

Q. 10 The average hydrogen bond length of O-H.....N is:

(*a*) 0.278 A° (*b*) 2.78 A°

(*c*) 27.8 A° (*d*) 278 A°

Q. 11 The average hydrogen bond length of N-H.....N is:

(*a*) 0.307 A° (*b*) 3.07 A°

(*c*) 30.7 A° (*d*) 307 A°

Q. 12 The average hydrogen bond length of N-H......S is:

(*a*) 0.340 A° (*b*) 3.40 A°

(*c*) 34.0 A° (*d*) 340 A°

Q. 13 The hydrogen bonds formed with C-H groups are .... When compared to the H bonds involving F, O, N, Cl, S etc.

(*a*) Very weak (*b*) Equal

(*c*) Strong (*d*) Very strong

Q. 14 In Coulomb's law, magnitude of the charge is:

(*a*) Directly proportional

(*b*) Inversely proportional

(*c*) Indirectly related

(*d*) Directly related

Q. 15 In Coulomb's law, square of the distance separating charged atoms are:

(*a*) Directly proportional

(*b*) Inversely proportional

(*c*) Indirectly related

(*d*) Directly related

Q. 16 The presence of H bonding is detected by:

(*a*) Infrared absorption spectra

(*b*) X-ray analysis

(*c*) Electro diffraction examination

(*d*) All

Q. 17 Is there any role of H bonds in establishing the structure of biopolymers:

(*a*) Yes (*b*) No

Q. 18 Dipole moment of water is:

(*a*) 1.85 D (*b*) 2.85 D

(*c*) 3.85 D (*d*) 4.85 D

Q. 19 The molecule H-O-H is .... in nature:

(*a*) Linear (*b*) Non-linear

(*c*) Both (*d*) None

Q. 20 The molecule H-O-H is .... in nature:

(*a*) Symmetrical (*b*) Non-symmetrical

(*c*) Both (*d*) None

Q. 21 Dipole moment of each O-H bond of H_2O is:

(*a*) 1.60 D (*b*) 1.85 D

(*c*) 2.60 D (*d*) 285 D

Q. 22 The angle between two O-H bonds of H_2O is:

(*a*) 104.5° (*b*) 90.5°

(*c*) 87.5° (*d*) 66.6°

Q. 23 Dipole moment can be measured by determining ....

(*a*) Resistance (*b*) Potential difference

(*c*) Dielectric constant (*d*) Current

Q. 24 Dielectric constant can be measured from the ratio of the electrical capacity of .....

(*a*) Condenser filled with substance to empty condenser

(*b*) Empty condenser to filled condenser with substance

(*c*) Both (*a*) and (*b*)

(*d*) None

Q. 25 Which of the following solvent(s) is/are non-polar:

(*a*) Benzene (*b*) CCl_4

(*c*) Both (*d*) None

Q. 26 Dipole moment of organic molecules help in understanding their property/properties such as:

(*a*) Configuration

(*b*) Cis-trans isomerism

(*c*) Hydrogen bonding

(*d*) Infrared & Raman spectra

(*e*) All

Q. 27 The formation of H bonds is an:

(*a*) Exothermic process (*b*) Endothermic process

(*c*) Both (*d*) None

Q. 28 Dissociation of H bond is an:

(*a*) Exothermic process (*b*) Endothermic process

(*c*) Both (*a*) and (*b*) (*d*) None

Q. 29 The attractive force which binds the H atom of one molecule with an electro negative atom of another molecule, is known as:

(*a*) Hydrogen bond (*b*) Proton bond

(*c*) Both (*a*) and (*b*) (*d*) Covalent bond

Q. 30 Dipole moment is a .... quantity:

(*a*) Vector

(*b*) Scalar

(*c*) Both

(*d*) Vary as per the situation

Q. 31 Dipole moment is indicated by an arrow running parallel to the line joining the charges and pointing towards the:

(*a*) Positive end
(*b*) Negative end
(*c*) Neutral end
(*d*) Both (*a*) and (*b*)

Q. 32 The degree of polarity depends on:

(*a*) Differences in the electronegativities of 2 elements of a molecule

(*b*) Differences in the electropositivities of 2 elements of a molecule

(*c*) Both (*a*) and (*b*)

(*d*) Neither (*a*) nor (*b*)

Q. 33 A stronger electric field produces higher dipole moments and the values are temperature ....

(*a*) Dependant
(*b*) Independent
(*c*) Sensitive
(*d*) None

Q. 34 Which is the following is Polysaccharide:

(*a*) Amylose

(*b*) Amylopectin

(*c*) Both (*a*) and (*b*)

(*d*) Starch

Q. 35 Hydrogen bonding may be:

(*a*) Intermolecular
(*b*) Intra molecular
(*c*) Both (*a*) and (*b*)
(*d*) None

Q. 36 Intermolecular hydrogen bonding changes which of the following in the structure of the participants in a system?

(*a*) Number

(*b*) Mass

(*c*) Shape

(*d*) Electronic structure

(*e*) All

Q. 37: Due to the intermolecular association, the effective molecular weight of a structure is increased leading to:

(*a*) Higher melting point and higher boiling point

(*b*) Lower melting point and lower boiling point

(*c*) Higher melting point and lower boiling point

(*d*) Lower melting point and higher boiling point

Q. 38 Intramolecular H bonding takes place when the molecules are in which of the following conformation?

(*a*) cis (*b*) gauche

(*c*) skew (*d*) All

Q. 39 The secondary structure of protein is entirely due to the H-bonding between the C-O and N-H groups of which of the following bonds?

(*a*) Hydrogen bond (*b*) Covalent bond

(*c*) Peptide bond (*d*) Coordinate bond

Q. 40 Every amino group forms a H-bond ...... A° long with the carbonyl group of the fourth residue down along with the chain.

(*a*) 0.272 (*b*) 2.72

(*c*) 27.2 (*d*) 272

Q. 41 H-bonding, taking place between the side chains of amino acids, is one of the forces responsible for the folding of which of the following into tertiary or quaternary structure of protein:

(*a*) α-helix (*b*) β-helix

(*c*) γ-helix (*d*) None

Q. 42 Depending on the nature of the side chain groups, how many groups of amino acids are recognized?

(*a*) 2 (*b*) 3

(*c*) 4 (*d*) 5

Q. 43 Depending on the nature of the side chain groups, which of the following amino acid(s) is/are recognized?

(*a*) Amino acids with H-bond donar and acceptor groups

(*b*) Amino acids with H-bond donar groups only

(*c*) Amino acids with H-bond acceptor groups only

(*d*) All

Q. 44 Which of the following is/are amino acid(s) with H bond donar and acceptor groups?

(*a*) Histidine (*b*) Lysine

(*c*) Threonine (*d*) All

Q. 45 Which of the following is/are amino acid(s) with donar groups only?

(*a*) Arginine (*b*) Tryptophan

(*c*) Cysteine (*d*) All

Q. 46 Amino acids are:

(*a*) Acidic (*b*) Basic

(*c*) Amphoteric (*d*) None

Q. 47 Proteins are covalently bonded linear polymers of how many different α-amino acids:

(*a*) 10 (*b*) 20

(*c*) 30 (*d*) 40

Q. 48 The covalent linkages formed between the amino group of one amino acid and the carboxyl group of another amino acid is known as:

(*a*) Peptide linkage (*b*) Covalent linkage

(*c*) Hydrogen linkage (*d*) Carboxyl linkage

Q. 49 The polymer of amino acids containing many peptide bonds are known as:

(*a*) Polypeptide (*b*) Multipeptide

(*c*) Covalent linkage (*d*) α-helix

Q. 50 One end of polypeptide of amino acid has a free amino group, which is said to be:

(*a*) C-terminal (*b*) N-terminal

(*c*) H-terminal (*d*) O-terminal

Q. 51 One end of polypeptide of amino acid has a free carboxylic group, which is said to be:

(*a*) C-terminal (*b*) N-terminal

(*c*) H-terminal (*d*) O-terminal

Q. 52 The peptide C-N bond length is .... than the normal C-N bond length

(*a*) more (*b*) shorter

(*c*) equal (*d*) none

Q. 53 Secondary structure of protein(s) include:

(*a*) alpha helix (*b*) Beta pleated sheet

(*c*) Collagen (*d*) All

Q. 54 The angle of rotation around the N-Cα bond is in the counter clockwise direction and is called:

(*a*) Phai (ϕ) (*b*) pSI (ψ)

(*c*) Alpha (α) (*d*) Beta (β)

Q. 55: The angle of rotation around the N-Cα_2 bond in the clockwise direction is called:

(*a*) Phai (ϕ) (*b*) pSI (ψ)

(*c*) Alpha (α) (*d*) Beta (β)

Q. 56 When ϕ and ψ are both 00, all atoms except the R groups and the hydrogens attached to the α-carbons lie in which of the following plane:

(*a*) same (*b*) opposite

(*c*) Reverse (*d*) Vertical

Q. 57 If ϕ versus ψ graph is plotted, the graph is called as:

(*a*) Ramachandran diagram

(*b*) Ramachandran graph

(*c*) Ramachandran sheet
(*d*) Ramachandran figure

Q. 58 Which of the following is/are the type of helix/helices in protein:

(*a*) Alpha helix (*b*) P ai helix
(*c*) $^{3}10$ helix (*d*) All

Q. 59 Majority of the alpha helix in protein is:

(*a*) Left handed (*b*) Right handed
(*c*) Vertical (*d*) Horizontal

Q. 60 If the alighnment of the C-terminus of one plolypeptide with the C-terminus of another is referred as:

(*a*) Parallel (*b*) Antiparallel
(*c*) Vertical up (*d*) Vertical down

Q. 61 The interaction found in tertiary structure protein is:
Van der Waals forces

(*a*) H bonds
(*b*) Electrostatic interaction
(*c*) Disulphide bonds
(*d*) All

Q. 62 The overall side chains of non-polar amino acids and residues to collect in the interior of the protein is known as:

(*a*) Hydrophillic effect (*b*) Hydrophobic effect
(*c*) Both (*d*) None

Q. 63 Which of the following is the correct of hydrophobic character of the amino acids:

(*a*) Cys > Leu > Gly > Phe > Val > Ala
(*b*) Arg > Gln > Asn > Try > Glu > Asp > Ser
(*c*) Phe > Ala > Val > Gly > Leu > Cys
(*d*) Try > Ser > Asp > Glu > Asn > Gln > Arg

Q. 64 Which of the following is the correct of hydrophillic character of the amino acids:

(*a*) Cys > Leu > Gly > Phe > Val > Ala

(*b*) Arg > Gln > Asn > Try > Glu > Asp > Ser

(*c*) Phe > Ala > Val > Gly > Leu > Cys

(*d*) Try > Ser > Asp > Glu > Asn > Gln > Arg

Q. 65 The hydrogen bond between side chains take place in the hydrophobic interior of proteins where competition of water is:

(*a*) Less (*b*) High

(*c*) Very high (*d*) Extremely high

Q. 66 The positively charged side chains in proteins get attracted towards each other because of electrostatic interactions, the bridges made is known as:

(*a*) Salt bridge (*b*) Electrostatic bridge

(*c*) Hydrogen bridge (*d*) Disulphide bridge

Q. 67 Which of the following bond is formed when the side chains of 2 cysteine residues come closer together resulting in the formation of a cysteins residues:

(*a*) Salt bridge (*b*) Electrostatic bridge

(*c*) Disulphide bridge (*d*) Hydrogen bridge

Q. 68 The properties of a quanternary protein is the–effect of the monomers and are mostly dependent on the molecular symmetry:

(*a*) Anagistic effect

(*b*) Synergistic effect

(*c*) Both (*a*) and (*b*)

(*d*) None

Q. 69 Nucleic acids are.............. and complex molecules which are responsible for the storage, transmission and translation of the genetic information:

(*a*) Extremely small (*b*) Large

(*c*) Extremely large (*d*) Small

Q. 70 Nucleic acids are copolymers of nucleotides that contains bases and a:

(*a*) Lupid (*b*) Protein

(*c*) Sugar (*d*) Amino acid

Q. 71 A unit of nucleic acid is a nucleotide is consisting of which of the following component:

(*a*) Sugar (*b*) Phosphoric acid

(*c*) Base (*d*) All

Q. 72 Which of the following nucleic acid is distinguished among the sugars:

(*a*) RNA (*b*) DNA

(*c*) Both (*a*) and (*b*) (*d*) None

Q. 73 DNA contains:

(*a*) Ribose (*b*) Deoxy ribose

(*c*) Both (*a*) and (*b*) (*d*) None

Q. 74 The heterocyclic phases found in the nucleotides are belonging to which of the following class:

(*a*) Purines (*b*) Pyrimidines

(*c*) Both (*a*) and (*b*) (*d*) None

Q. 75 How many N atoms are available in a single pyrimidine:

(*a*) 2 (*b*) 3

(*c*) 4 (*d*) 5

(*e*) 6

Q. 76 At which position N is attached in Pyrimidine:

(*a*) One and two (*b*) One and three

(*c*) One and four (*d*) One and five

Q. 77 Uracil found in:

(*a*) RNA (*b*) DNA

(*c*) Both (*a*) and (*b*) (*d*) NONE

Q. 78 Cytocine is found in:

(*a*) RNA (*b*) DNA

(*c*) Both (*a*) and (*b*) (*d*) NONE

Q. 79 Thymine is found in:

(*a*) RNA (*b*) DNA

(*c*) Both (*a*) and (*b*) (*d*) NONE

Q. 80 How many 5-methyl cytocine and 5-hydroxy methyl cytosine bases are found in DNA/RNA:

(*a*) Very less

(*b*) Large

(*c*) Very large

(*d*) Only both bases are available

Q. 81 Which of the following disrupt DNA replication while undergoing photodimerization in the presence of UV light:

(*a*) Thymine (*b*) Uracil

(*c*) Cytosine (*d*) Both (*b*) and (*c*)

Q. 82 Which of the following is a purine:

(*a*) Adenin (*b*) Guanine

(*c*) Both (*a*) and (*b*) (*d*) None

Q. 83 Which of the following purines is found as important metabolite:

(*a*) Xanthyne (*b*) Hypoxanthyne

(*c*) Uric acid (*d*) All

Q. 84 Sugar phosphate ester of the nucleosides are known as:

(*a*) Nucleotide (*b*) Heterocyclic bases

(*c*) DNA (*d*) RNA

Q. 85 Length of DNA varies from:

(*a*) 1 mm to 50mm (*b*) 1mm to 60mm

(*c*) 1mm to 70mm (*d*) 1mm to 80mm

Q. 86 Which of the following chromosomes have longest DNA:

(*a*) Viruses (*b*) Bacteria

(*c*) Human (*d*) Equal in all

Q. 87 Secondary structure of DNA represents the duplex formation and further ____ of thee DNA:

(*a*) Coiling (*b*) Parallel
(*c*) Uncoiling (*d*) Vertical

Q. 88 The duplex formation of DNA is due to the hydrogen bonds developed between:

(*a*) Acids (*b*) Bases
(*c*) Alkali (*d*) Both (*a*) and (*c*)

Q. 89 Adenine of one chain always pairs with:

(*a*) Thymine (*b*) Cytocine
(*c*) Guanine (*d*) Adenine

Q. 90 Guanine of first chain pairs which of the following second chain:

(*a*) Thymine (*b*) Cytocine
(*c*) Guanine (*d*) Adenine

Q. 91 Double stranded molecules get iled about a central axis and produces a right handed:

(*a*) double helix (*b*) Triple helix
(*c*) Single helix (*d*) Both (*a*) and (*b*)

Q. 92 The attraction between base pairs causes the twisting of two strands of the double helix:

(*a*) Straight (*b*) Opposite parallel
(*c*) Both (*a*) and (*c*) (*d*) None

Q. 93 Adjacent base pairs in the double helix are rotated by:

(*a*) 26 degree (*b*) 32 degree
(*c*) 36 degree (*d*) 39 degree
(*e*) 42 degree

Q. 94 The double helix completes one turn every-residues:

(*a*) 6 (*b*) 8
(*c*) 10 (*d*) 12

Q. 95 In double helix structure of DNA one turn advances the helix by:

(*a*) 0.34nm (*b*) 3.4nm

(*c*) 34nm (*d*) 340nm

Q. 96 In double helix structure each base pair advances the double helix by:

(*a*) 0.34nm (*b*) 3.4nm

(*c*) 34nm (*d*) 340nm

Q. 97 In double helix structure of DNA the bases are closely packed in the interior of the duplex at:

(*a*) Vander Waals distance (*b*) Covalent distance

(*c*) Coordinate distance (*d*) None

Q. 98 The standard free energy of formation of adenine, thymine base pair is:

(*a*) 4.0 KJ/mole (*b*) 5.0 KJ/mole

(*c*) 6.0 KJ/mole (*d*) 7.0 KJ/mole

Q. 99 Double helix has a dia of:

(*a*) 1 nm (*b*) 2 nm

(*c*) 3 nm (*d*) 4 nm

Q. 100 The functional processes of a DNA molecule cause unwinding and rewinding of the DNA helix and these windings produce a:

(*a*) Supercoiled form (*b*) Coiled form

(*c*) Straight form (*d*) None

Q. 101 The twisting of the double helix around its own axis is Right handed, the DNA is said to be:

(*a*) Positively supercoiled

(*b*) Negatively supercoiled

(*c*) Both (*a*) and (*b*)

(*d*) None

Q. 102 The twisting of the double helix around its own axis is left handed, the DNA is said to be:

(*a*) Positively supercoiled

(*b*) Negatively supercoiled

(*c*) Both (*a*) and (*b*)

(*d*) None

Q. 103 Supercoiling of DNA is quantified by:

(*a*) Winding number (*b*) Super helix density

(*c*) Both (*a*) and (*b*) (*d*) None

Q. 104 Super helix density of DNA is expressed by:

(*a*) No. of superhelical terms per base pairs

(*b*) Base pairs/Superhelical terms

(*c*) Binary numbers/Base pairs

(*d*) Base pairs/Winding number

Q. 105 The super helix density of a typical DNA molecule is having how many terms/1000 base pairs:

(*a*) 3 (*b*) 5

(*c*) 7 (*d*) 9

Q. 106 Ribonucleotide residues are interlinked by which of the following phospho diester bond:

(*a*) 3′–5′ (*b*) 2′–5′

(*c*) 1′–5′ (*d*) 2′–4′

Q. 107 Ribosomal RNA amounts to –% of the cellular component of RNA:

(*a*) 25% (*b*) 50%

(*c*) 75% (*d*) Nil

Q. 108 In amylose, glucose residues are linked by

(*a*) $\propto (1 \rightarrow 4)$ (*b*) $\propto (1 \rightarrow 6)$

(*c*) Both (*a*) and (*b*) (*d*) $\propto (2 \rightarrow 4)$

Q. 109 In amylopectin, glucose residues are linked by

(*a*) $\propto (1 \rightarrow 4)$ (*b*) $\propto (1 \rightarrow 6)$

(*c*) Both (*a*) and (*b*) (*d*) $\propto (2 \rightarrow 4)$

Q. 110 Cellulose is the most abundant polysaccharide in plants and is a linear polymer of glucose residues linked by

(*a*) $\beta (1 \rightarrow 4)$ (*b*) $\beta (1 \rightarrow 6)$

(*c*) $\beta (2 \rightarrow 4)$ (*d*) $\beta (2 \rightarrow 6)$

Answeres

1	*(a)*	23	*(c)*	45	*(d)*	67	*(c)*
2	*(b)*	24	*(c)*	46	*(c)*	68	*(b)*
3	*(d)*	25	*(c)*	47	*(b)*	69	*(c)*
4	*(a)*	26	*(e)*	48	*(a)*	70	*(c)*
5	*(c)*	27	*(a)*	49	*(a)*	71	*(d)*
6	*(a)*	28	*(b)*	50	*(b)*	72	*(c)*
7	*(b)*	29	*(c)*	51	*(a)*	73	*(b)*
8	*(b)*	30	*(a)*	52	*(b)*	74	*(c)*
9	*(b)*	31	*(b)*	53	*(d)*	75	*(a)*
10	*(b)*	32	*(a)*	54	*(a)*	76	*(b)*
11	*(b)*	33	*(b)*	55	*(b)*	77	*(a)*
12	*(b)*	34	*(c)*	56	*(a)*	78	*(c)*
13	*(a)*	35	*(c)*	57	*(a)*	79	*(b)*
14	*(a)*	36	*(e)*	58	*(d)*	80	*(a)*
15	*(b)*	37	*(a)*	59	*(b)*	81	*(a)*
16	*(d)*	38	*(d)*	60	*(a)*	82	*(b)*
17	*(a)*	39	*(c)*	61	*(e)*	83	*(d)*
18	*(a)*	40	*(b)*	62	*(b)*	84	*(a)*
19	*(b)*	41	*(a)*	63	*(c)*	85	*(d)*
20	*(b)*	42	*(b)*	64	*(d)*	86	*(c)*
21	*(a)*	43	*(d)*	65	*(a)*	87	*(a)*
22	*(a)*	44	*(d)*	66	*(a)*	88	*(b)*

89	**(*a*)**	**95**	**(*b*)**	**101**	**(*a*)**	**107**	**(*c*)**
90	**(*b*)**	**96**	**(*a*)**	**102**	**(*b*)**	**108**	**(*a*)**
91	**(*a*)**	**97**	**(*a*)**	**103**	**(*c*)**	**109**	**(*c*)**
92	**(*b*)**	**98**	**(*b*)**	**104**	**(*a*)**	**110**	**(*a*)**
93	**(*c*)**	**99**	**(*b*)**	**105**	**(*b*)**		
94	**(*c*)**	**100**	**(*a*)**	**106**	**(*a*)**		

Chapter 3
Nuclear Physics

Q. 1 The clue for the existence of the particles in the nucleus of an atom was given by:

(*a*) Discharge tube experiment

(*b*) X-ray analysis

(*c*) Positive ray analysis

(*d*) Alpha ray scattering experiment

Q. 2 In which of the following n/p ratio is greater than one?

(*a*) Na^+ (*b*) Mg^{2+}

(*c*) O^{2-} (*d*) Al^{3+}

Q. 3 Unstability of the nucleus is due to:

(*a*) High neutron/proton ratio

(*b*) Low neutron/proton ratio

(*c*) High electron/proton ratio

(*d*) High proton/electron ratio

Q. 4 Transition of an electron from M level to K level results in:

(*a*) Absorption spectra (*b*) Emission spectra

(*c*) X-ray spectra (*d*) Continuous spectra

Q. 5 The positron has a charge equal to that of:

(*a*) A proton (*b*) An electron

(*c*) An α-particle (*d*) A neutron

Q. 6 The electronic energy levels of the hydrogen atom in the Bohr theory are called as:

(*a*) Orbitals (*b*) Orbits

(*c*) Rydberg levels (*d*) Ground states

Q. 7 A node is a surface on which the probability of finding an electron is:

(*a*) 0 (*b*) > 1

(*c*) > 10 (*d*) > 90

Q. 8 Bohr's model of an atom is based on:

(*a*) Quantum theory

(*b*) Dalton atomic theory

(*c*) Theory of electrolytic dissociation

(*d*) Law of mass action

Q. 9 The statement "orbitals of equivalent energy are filled with electrons before spin pairing occurs" is called:

(*a*) Uncertainty principle

(*b*) Paulis exclusion principle

(*c*) Zeeman effect

(*d*) Hund's rule

Q. 10 "The principle that no two electrons can have all the four quantum number is identical" was enunciated by:

(*a*) Davy (*b*) Rutherford

(*c*) Pauli (*d*) Einstein

(*e*) Seaborg

Q. 11 Orbital number of an atom also indicates its possible number of:

(*a*) Electrons (*b*) Protons

(*c*) Orbitals (*d*) Neutrons

Q. 12 The third orbit of any atom of an element contains:

(*a*) 3 electrons only (*b*) 3 protons only

(*c*) 3 neutrons only (*d*) 3 orbitals only

Q. 13 The Aufbau principle implies that a new electron occupies an orbital for which of the following?

(*a*) n has a lower value

(*b*) *l* has a lower value

(*c*) The value (n+l) is minimum

(*d*) The value (n+l) is maximum

Q. 14 The spectrum of He is expected to be similar to that of:

(*a*) H (*b*) Li^+

(*c*) Na (*d*) He^+

Q. 15 For a hydrogen atom $E_n = -13.6/n^2$ Ev/atom. Therefore minimum energy that an atom of hydrogen in the ground state can absorb is:

(*a*) 13.6 eV (*b*) 10.2 eV

(*c*) 3.4 eV (*d*) 1.5 eV

Q. 16 Alpha rays are electrically:

(*a*) Positively charged

(*b*) Negatively charged

(*c*) Neutral

(*d*) None

Q. 17 Radioactive substances emit:

(*a*) Hydrogen (*b*) Helium

(*c*) Nitrogen (*d*) α, β and γ-rays

Q. 18 Natural radioactivity was discovered by:

(*a*) Rutherford

(*b*) Henri Becquerel

(*c*) Madame Curie

(*d*) Schmidt

Q. 19 Which of the following has the greatest penetrating power?

(*a*) Alpha rays (*b*) Beta rays

(*c*) Gamma rays (*d*) None

Q. 20 In a beta ray change:

(*a*) An electron from the extra-nuclear electrons of the atom is given out

(*b*) An electron from the nucleus is given out

(*c*) A proton from the nucleus is given out

(*d*) A neutron from the nucleus is given out

Q. 21 When a radioactive substance is subjected to a vacuum, rate of disintegration per second:

(*a*) Increases considerably

(*b*) Increases only if the products are gaseous

(*c*) Is not affected

(*d*) Suffers a slight decrease

Q. 22 The half life of radium is 1600 years. After how much time it will remain 25% of the present?

(*a*) 8000 years (*b*) 1600 years

(*c*) 3200 years (*d*) 4800 years

Q. 23 When the quantum of radioactive substance is increased two times, the number of atoms disintegrating per unit time is:

(*a*) Doubled

(*b*) Increased by square of two

(*c*) Increased but not to a great extent

(*d*) Not affected

Q. 24 A radioactive atom emits a beta ray (an electron) from:

(*a*) Its outermost orbit

(*b*) Its nucleus

(*c*) Its innermost orbit

(*d*) None

Q. 25 The emission of beta rays produces an:

(*a*) Isotope (*b*) Isobar

(*c*) Isotone (*d*) All

Q. 26 Which of the following is the isotone of $_{16}S^{32}$?

(*a*) $_{15}P^{31}$ (*b*) $_{8}O^{16}$

(*c*) $_{17}Cl^{35}$ (*d*) $_{6}C^{12}$

Q. 27 Alpha rays are:

(*a*) He^{+} ions (*b*) He^{+} ions

(*c*) He atoms (*d*) Electrons

(*e*) Fast moving electrons

Q. 28 Isotope differs in:

(*a*) The number of protons

(*b*) The valence electrons

(*c*) The chemical activity

(*d*) The number of neutrons

Q. 29 Elements having different nuclear charge but the same atomic mass are called as:

(*a*) Isotope (*b*) Isobars

(*c*) Isotone (*d*) Isomers

Q. 30 The element which is naturally radioactive is:

(*a*) Nitrogen

(*b*) Oxygen

(*c*) Phosphorus

(*d*) Radium

(*e*) Barium

Q. 31 Atomic number and atomic mass of Uranium is 92 and 238. During the course of radioactive disintegration it loses 5 α-particles and 4 β-particles. The atomic number and atomic weight of the new element will be:

(*a*) At. no. = 90, at. wt. = 233

(*b*) At. no. = 86, at. wt. = 218

(*c*) At. no. = 82, at. wt. = 218

(*d*) At. no. = 88, at. wt. = 228

Q. 32 Atomic weight of thorium is 232 and the atomic number is 90. The end product of its disintegration is an isotope of lead (At. wt.=208, At.no.=82). The number of α and β-particles emitted is

(*a*) $\alpha = 3; \beta = 3$ (*b*) $\alpha = 6; \beta = 4$

(*c*) $\alpha = 6; \beta = 0$ (*d*) $\alpha = 4; \beta = 6$

Q. 33 After 2 hours the amount of a radioactive substance is reduced to 1/16 of its original amount. The half-life of the radioactive substance is:

(*a*) 15 minutes (*b*) 30 minutes

(*c*) 45 minutes (*d*) 60 minutes

Q. 34 A beta ray change results in the production of:

(*a*) An isotope (*b*) An isobar

(*c*) An allotrope (*d*) An isomer

Q. 35 The first artificial transmutation of elements was carried out by:

(*a*) Madame Curie (*b*) Irene Curie

(*c*) Rutherford (*d*) Soddy

Q. 36 Artificial radioactive was discovered by:

(*a*) Madame Curie (*b*) Fermi

(*c*) Irene Curie & Federic Joliot (*d*) Rutherford

Q. 37 If radium and chlorine combines to form radium chloride, the compound is:

(*a*) No longer radioactive

(*b*) Half as radioactive as the radium content

(*c*) As radioactive as the radium content

(*d*) Twice as radioactive as the radium content

Q. 38 Which of the following statements about radioactivity of an element is incorrect?

(*a*) It is a nuclear property

(*b*) It does not involve any rearrangement of electrons

(*c*) Its rate affected by change in temperature and pressure

(*d*) It remains unaffected by the presence of other element chemically combined with it

Q. 39 The half-life of a radioactive element is 35 years. If there are 4×10^6 nuclei at the start, then after how many years they will be left 0.5×10^6?

(*a*) 35 (*b*) 70

(*c*) 105 (*d*) 140

Q. 40 If 4 gm. of a radioactive isotope has half-life of 10 days, the half-life of 2 gm. sample is:

(*a*) 10 days (*b*) 20 days

(*c*) 30 days (*d*) 5 days

Q. 41 The atomic weight of uranium is 238 while its atomic no. is 92. If it emits 8α and 6β particles, then the new element will have

(*a*) At. wt. 206 and at. no. 82

(*b*) At. wt. 206 and at. no. 76

(*c*) At. wt. 224 and at. no. 80

(*d*) At. wt. 230 and at. no. 90

Q. 42 The emission of a β ray produces:

(*a*) Isobar (*b*) Isotope

(*c*) Isotone (*d*) Isomer

Q. 43 α-rays consist of a stream of:

(*a*) H (*b*) He^+ ions

(*c*) Only electrons (*d*) Only neutrons

Q. 44 Tritium is an isotope of:

(*a*) Tellurium (*b*) Tritanium

(*c*) Tentalum (*d*) Hydrogen

Q. 45 A device used for the measurement of radioactivity is a:

(*a*) Mass spectrometer (*b*) Cyclotron

(*c*) Nuclear reactor (*d*) G. M. counters

Q. 46 Isotopes are separated by:

(*a*) Physical property (*b*) Mass

(*c*) Both (*d*) None

Q. 47 Detection and measurement of radioactivity involves which of the following methods:

(*a*) Cloud chamber

(*b*) Geiger-Muller Counter

(*c*) Film Badges

(*d*) All of the above

Q. 48 1 Rutherford is equal to how many disintegration per second:

(*a*) 10^2 (*b*) 10^4

(*c*) 10^5 (*d*) 10^6

Q. 49 Which of the following is the SI unit of radioactivity:

(*a*) Rutherford (*b*) Curie

(*c*) Becquereal (*d*) None

Q. 50 When a pair of nuclei having same number of protons and neutrons but different half lives, the phenomenon is called as:

(*a*) Nuclear isomerism (*b*) Nuclear polymerism

(*c*) Nuclear isomorphism (*d*) None

ANSWERS

1 (*d*)	**11 (*d*)**	**21 (*c*)**	**31 (*c*)**	**41 (*a*)**
2 (*c*)	**12 (*c*)**	**22 (*a*)**	**32 (*b*)**	**42 (*a*)**
3 (*a*)	**13 (*b*)**	**23 (*a*)**	**33 (*b*)**	**43 (*b*)**
4 (*b*)	**14 (*a*)**	**24 (*b*)**	**34 (*a*)**	**44 (*d*)**
5 (*a*)	**15 (*a*)**	**25 (*a*)**	**35 (*a*)**	**45 (*d*)**
6 (*b*)	**16 (*d*)**	**26 (*d*)**	**36 (*c*)**	**46 (*c*)**
7 (*a*)	**17 (*b*)**	**27 (*d*)**	**37 (*c*)**	**47 (*d*)**
8 (*d*)	**18 (*b*)**	**28 (*b*)**	**38 (*b*)**	**48 (*d*)**
9 (*c*)	**19 (*a*)**	**29 (*d*)**	**39 (*c*)**	**49 (*c*)**
10 (*c*)	**20 (*c*)**	**30 (*b*)**	**40 (*a*)**	**50 (*a*)**

Chapter 4
Agrometeorology

Q. 1 The condition of atmosphere at a given place and time is termed as:

(*a*) Climate (*b*) Weather

(*c*) Meteorology (*d*) None

Q. 2 Weather pertains to a:

(*a*) Very long period of time (*b*) small duration

(*c*) medium duration (*d*) None

Q. 3 Climate pertains to a:

(*a*) long period of time (*b*) small duration

(*c*) medium duration (*d*) None

Q. 4 Weather pertains to a:

(*a*) long period of time (*b*) small duration

(*c*) medium duration (*d*) None

Q. 5 The summation of weather conditions over a given region during a comparatively longer period is termed as:

(*a*) Climate (*b*) Weather

(*c*) Meteorology (*d*) None

Q. 6 The science which deals with laws and principles of atmospheric phenomenon:

(*a*) Climatology

(*b*) Agricultural meteorology

(*c*) Meteorology

(*d*) None

Q. 7 The branch of meteorology that deals with the response of crops to the physical environment:

(*a*) Climatology

(*b*) Agricultural meteorology

(*c*) Agro climatology

(*d*) None

Q. 8 The relationship of climatic regimes and agricultural production is termed as:

(*a*) Climatology

(*b*) Agricultural meteorology

(*c*) Agro climatology

(*d*) None

Q. 9 The gaseous envelop of invisible film of air surrounding the earth is called:

(*a*) Climate (*b*) Weather

(*c*) Troposphere (*d*) Atmosphere

Q. 10 Atmosphere extends up to a height of:

(*a*) 600 km (*b*) 60 km

(*c*) 6000 km (*d*) 1600 km

Q. 11 Air contains:

(*a*) 21% Nitrogen (*b*) 87% Nitrogen

(*c*) 78% Nitrogen (*d*) 0.03% Nitrogen

Q. 12 Air contains:

(*a*) 21% Oxygen (*b*) 87% Oxygen

(*c*) 78% Oxygen (*d*) 0.03% Oxygen

Q. 13 Air contains:

(*a*) 21% CO_2 (*b*) 87% CO_2

(*c*) 78% CO_2 (*d*) 0.03% CO_2

Q. 14 Air contains:

(*a*) 2.1% Argon (*b*) 0.93% Argon

(*c*) 7.8% Argon (*d*) 0.03% Argon

Q. 15 Which is not a major component of atmosphere

(a) Carbon dioxide
(b) Oxygen
(c) Argon
(d) Carbon monoxide

Q. 16 Based on vertical temperature differences atmosphere is divided into:

(a) 4 layers
(b) 3 layers
(c) 5 layers
(d) 6 layers

Q. 17 The lower most layer of atmosphere is:

(a) Troposphere
(b) Stratosphere
(c) Mesosphere
(d) Thermosphere

Q. 18 The upper most layer of atmosphere is:

(a) Troposphere
(b) Stratosphere
(c) Mesosphere
(d) Thermosphere

Q. 19 The second layer of atmosphere is:

(a) Troposphere
(b) Stratosphere
(c) Mesosphere
(d) Thermosphere

Q. 20 The third layer of atmosphere is:

(a) Troposphere
(b) Stratosphere
(c) Mesosphere
(d) Thermosphere

Q. 21 The densest part of the atmosphere is:

(a) Troposphere
(b) Stratosphere
(c) Mesosphere
(d) Thermosphere

Q. 22 Troposphere is thicker at:

(a) Poles
(b) Equator
(c) Both
(d) None

Q. 23 Troposphere extends up to a height of:

(a) 8-18 km
(b) 18-50 km
(c) 50-80 km
(d) >80 km

Q. 24 Stratosphere extends up to a height of:

(*a*) 8-18 km (*b*) 18-50 km

(*c*) 50-80 km (*d*) > 80 km

Q. 25 Mesosphere extends up to a height of:

(*a*) 8-18 km (*b*) 18-50 km

(*c*) 50-80 km (*d*) > 80 km

Q. 26 Thermosphere extends up to a height of:

(*a*) 8-18 km (*b*) 18-50 km

(*c*) 50-80 km (*d*) > 80 km

Q. 27 In troposphere the temperature:

(*a*) Increases with altitude

(*b*) Decreases with altitude

(*c*) Remains constant

(*d*) None

Q. 28 The temperature in the boundary of troposphere is:

(*a*) 60°C (*b*) –60°C

(*c*) 6°C (*d*) –6°C

Q. 29 The thin layer separating stratosphere from troposphere is:

(*a*) Tropopause (*b*) Stratopause

(*c*) Mesopause (*d*) Thermopause

Q. 30 The thin layer separating stratosphere from mesosphere is:

(*a*) Tropopause (*b*) Stratopause

(*c*) Mesopause (*d*) Thermopause

Q. 31 The thin layer separating mesosphere from thermosphere is:

(*a*) Tropopause (*b*) Stratopause

(*c*) Mesopause (*d*) Thermopause

Q. 32 The term ozone layer is synonymous with:

(*a*) Tropopause (*b*) Stratopause

(*c*) Mesopause (*d*) Thermopause

Q. 33 In thermosphere the temperature:

(*a*) Steadily increases with altitude

(*b*) Steadily decreases with altitude

(*c*) Remains constant

(*d*) None

Q. 34 The amount of solar radiation continuously emitted by sun is:

(*a*) 1.94 cal/cm/min (*b*) 1.94 cal/cm^2/min

(*c*) 1.94 cal/cm^2/hr (*d*) 1.94 cal/cm^2/day

Q. 35 Solar radiation is received in the form of:

(*a*) Electromagnetic waves (*b*) Magnetic waves

(*c*) Thermo magnetic waves (*d*) None

Q. 36 Photo synthetically active radiation (PAR) is more in:

(*a*) Direct radiation (*b*) Diffused radiation

(*c*) equal (*d*) None

Q. 37 The solar radiation reflected without any change in its quality is called:

(*a*) Torpedo (*b*) Albedo

(*c*) Armada (*d*) None

Q. 38 Thermal radiation emitted by the earth is termed as:

(*a*) Solar radiation (*b*) Lunar radiation

(*c*) Terrestrial radiation (*d*) None

Q. 39 The radiation balance between global and reflected solar radiation is called:

(*a*) Total radiation (*b*) Gross radiation

(*c*) Net radiation (*d*) None

Q. 40 The instrument used for measurement of total incoming solar radiation is:

(*a*) Solar meter (*b*) Pyranometer

(*c*) Thermometer (*d*) None

Q. 41 The instrument used for measurement of reflectivity:

(*a*) Torpedo meter (*b*) Albedo meter

(*c*) Armada meter (*d*) None

Q. 42 Photo synthetically active radiation (PAR) is measured with:

(*a*) Thermo sensor (*b*) Albedo meter

(*c*) Quantum sensor (*d*) None

Q. 43 Direct solar radiation is measured with:

(*a*) Pyrenometer (*b*) Pyrheliometer

(*c*) Polymeter (*d*) None

Q. 44 Heat can flow between substances through:

(*a*) Conduction (*b*) Convection

(*c*) Radiation (*d*) All

Q. 45 The transfer of heat by molecular activity is called:

(*a*) Conduction (*b*) Convection

(*c*) Radiation (*d*) All

Q. 46 The process of heat transfer within liquids and gases resulting from the motion of fluids is called:

(*a*) Conduction (*b*) Convection

(*c*) Radiation (*d*) All

Q. 47 The transfer of energy by electromagnetic waves is called:

(*a*) Conduction (*b*) Convection

(*c*) Radiation (*d*) All

Q. 48 Conduction is:

(*a*) The transfer of heat by molecular activity

(*b*) The process of heat transfer within liquids and gases resulting from the motion of fluids (*c*) The transfer of energy by electromagnetic waves

(*d*) All

Q. 49 Convection is:

(*a*) The transfer of heat by molecular activity

(*b*) The process of heat transfer within liquids and gases resulting from the motion of fluids

(*c*) The transfer of energy by electromagnetic waves

(*d*) All

Q. 50 Radiation is:

(*a*) The transfer of heat by molecular activity

(*b*) The process of heat transfer within liquids and gases resulting from the motion of fluids

(*c*) The transfer of energy by electromagnetic waves

(*d*) All

Q. 51 Which of the following does not require the presence of medium:

(*a*) Conduction (*b*) Convection

(*c*) Radiation (*d*) All

Q. 52 The temperature variation between day and night is called:

(*a*) Diurnal variation (*b*) Seasonal variation

(*c*) Both (*a*) and (*b*) (*d*) Bone

Q. 53 The temperature variation between seasons is called:

(*a*) Diurnal variation (*b*) Seasonal variation

(*c*) Both (*a*) and (*b*) (*d*) Bone

Q. 54 The diurnal variation in temperature is less at:

(*a*) Coastal areas (*b*) Equator

(*c*) Inlands (*d*) None

Q. 55 The Seasonal variation in temperature is less at:

(*a*) Coastal areas (*b*) Equator

(*c*) Inlands (*d*) Higher altitudes

Q. 56 The lines joining the points of equal temperature:

(*a*) Isobytes (*b*) Isotherms

(*c*) Isobar (*d*) None

Q. 57 The decrease in temperature in higher altitudes in the air is called:

(*a*) Horizontal temperature gradient

(*b*) Vertical temperature gradient

(*c*) Isotherm

(*d*) None

Q. 58 The vertical temperature gradient is expressed as:

(*a*) Floating rate (*b*) Lapse rate

(*c*) Print rate (*d*) Pulse rate

Q. 59 The normal lapse rate is:

(*a*) 6.5°C/km (*b*) 65°C/km

(*c*) –6.5°C/km (*d*) –65°C/km

Q. 60 The rate at which temperature changes as air rises or falls is called:

(*a*) Actual lapse rate (*b*) Adiabatic lapse rate

(*c*) Thermal lapse rate (*d*) None

Q. 61 Inversion is the:

(*a*) Abrupt fall instead of rise in temperature

(*b*) Vertical temperature gradient

(*c*) Abrupt rise instead of fall in temperature

(*d*) None

Q. 62 The heat flow into or out of the soil is called:

(*a*) Actual lapse rate (*b*) Solar heat flux

(*c*) Thermal lapse rate (*d*) Soil heat flux

Q. 63 The thermal conductivity into the soil depends on:

(*a*) Texture (*b*) Moisture

(*c*) Organic matter content (*d*) All

Q. 64 Which of the following can influence the heat flow in soil:

(*a*) Irrigation (*b*) Ploughing

(*c*) Both (*d*) None

Q. 65 The melting and boiling point of water are:

(*a*) 32 and 210°F (*b*) 32 and 110°F

(*c*) 3.2 and 210°F (*d*) –32 and 210°F

Q. 66 The melting point of water is:

(*a*) 273°K (*b*) 100°K

(*c*) 0°K (*d*) –273°K

Q. 67 There is no activity of molecules of a substance at:

(*a*) 273°K (*b*) 100°K

(*c*) 0°K (*d*) "273°K

Q. 68 A mechanical device to record air temperature continuously:

(*a*) Thermometer (*b*) Thermograph

(*c*) Thermosensor (*d*) None

Q. 69 The pressure exerted by the atmosphere on the earth's surface is called:

(*a*) Thermal pressure

(*b*) Atmospheric pressure

(*c*) Solar pressure

(*d*) Vapour pressure

Q. 70 Pressure is:

(*a*) Weight/unit area (*b*) Force/unit area

(*c*) Volume/unit area (*d*) None

Q. 71 The atmospheric pressure:

(*a*) Decreases with decrease in elevation

(*b*) Decreases with increase in elevation

(*c*) Remains constant

(*d*) Increases with increase in elevation

72 Barometer is used to measure:

(*a*) Temperature (*b*) Wind

(*c*) Atmospheric pressure (*d*) None

Q. 73 The atmospheric pressure is expressed in:

(*a*) Bars (*b*) Millibars

(*c*) Centibars (*d*) None

Q. 74 Wind is air in horizontal motion which travels from a:

(*a*) High pressure area to a low pressure area

(*b*) Low pressure area to a high pressure area

(*c*) High tension area to a low tension area

(*d*) High tension area to a low tension area

Q. 75 The direction from which the wind comes is:

(*a*) Leeward (*b*) Windward

(*c*) Seaward (*d*) None

Q. 76 The direction towards which the wind blows is:

(*a*) Leeward (*b*) Windward

(*c*) Seaward (*d*) None

Q. 77 Wind direction is indicated by:

(*a*) Barometer (*b*) Wind vane

(*c*) Anemometer (*d*) None

Q. 78 Wind velocity is measured by:

(*a*) Barometer (*b*) Wind vane

(*c*) Anemometer (*d*) None

Q. 79 Wind velocity near the ground surface is:

(*a*) High than at higher elevations

(*b*) Less than at higher elevations

(*c*) Both

(*d*) None

Q. 80 Humidity refers to the:

(*a*) Temperature of the atmosphere

(*b*) Water vapour content of the atmosphere

(*c*) Gaseous content of the atmosphere

(*d*) Bone

Q. 81 The amount of water vapour in the atmosphere depends upon:

(*a*) Air and pressure (*b*) Air and temperature

(*c*) Pressure and temperature (*d*) None

Q. 82 When air contains maximum amount of water at a particular temperature it is said to be:

(*a*) Saturated (*b*) Unsaturated

(*c*) Both (*d*) None

Q. 83 The quantity of water vapour present in the atmosphere at any particular time is expressed as:

(*a*) Absolute humidity (*b*) Specific humidity

(*c*) Relative humidity (*d*) All

Q. 84 Absolute humidity is:

(*a*) Weight of water vapour/volume of air

(*b*) Weight of water vapour/weight of air including water vapour

(*c*) Water vapour present in the atmosphere/water vapour required for saturation × 100

(*d*) None

Q. 85 Specific humidity is:

(*a*) Weight of water vapour/volume of air

(*b*) Weight of water vapour/weight of air including water vapour

(*c*) Water vapour present in the atmosphere/water vapour required for saturation × 100

(*d*) None

Q. 86 Relative humidity is:

(*a*) Weight of water vapour/volume of air

(*b*) Weight of water vapour/weight of air including water vapour

(*c*) Water vapour present in the atmosphere/water vapour required for saturation × 100

(*d*) None

Q. 87 Absolute humidity is expressed in:

(*a*) g/m^2 (*b*) g/kg

(*c*) % (*d*) g/cc

Q. 88 Specific humidity is expressed in:

(*a*) g/m^2 (*b*) g/kg

(*c*) % (*d*) g/cc

Q. 89 Relative humidity is expressed in:

(*a*) g/m^2 (*b*) g/kg

(*c*) % (*d*) g/cc

Q. 90 Vapour pressure deficit is the difference between the:

(*a*) Saturation vapour pressure and unsaturated vapour pressure

(*b*) Actual vapour pressure and saturation vapour pressure

(*c*) Saturation vapour pressure and actual vapour pressure

(*d*) None

Q. 91 kPa stands for:

(*a*) Kelvin Pascals (*b*) Kelvin Pattern

(*c*) Kilo Pascals (*d*) Kilo Pattern

Q. 92 Air is said to be humid when the vapour pressure deficit is:

(*a*) <1.5kPa (*b*) >1.5kPa

(*c*) <2.5kPa (*d*) >2.5kPa

Q. 93 Air is said to be dry when the vapour pressure deficit is:

(*a*) <1.5kPa (*b*) >1.5kPa

(*c*) <2.5kPa (*d*) >2.5kPa

Q. 94 The instrument used for measuring relative humidity is:

(*a*) Psychrometer (*b*) Hygrometer

(*c*) Both (*a*) and (*b*) (*d*) None

Q. 95 The wet and dry bulb thermometers are kept in:

(*a*) Side screen (*b*) Sight screen

(*c*) Stevenson's screen (*d*) Sun screen

Q. 96 RH can be accurately measured in crop canopy with:

(*a*) Assman's psychrometer (*b*) Hair hygrometer

(*c*) Hygrograph (*d*) none

Q. 97 RH can be accurately measured inside the room with:

(*a*) Assman's psychrometer (*b*) Hair hygrometer

(*c*) Hygrograph (*d*) none

Q. 98 Humidity can be continuously measured with:

(*a*) Assman's psychrometer (*b*) Hair hygrometer

(*c*) Hygrograph (*d*) none

Q. 99 The sensitive element in hair hygrometer:

(*a*) Horse hair (*b*) Cattle hair

(*c*) Human hair (*d*) Thermometer

Q. 100 The process by which water in a liquid state is converted to vapour and escapes into air:

(*a*) Evaporation (*b*) Transpiration

(*c*) Evapotranspiration (*d*) Vapourisation

Q. 101 The equipment used for measurement of evaporation:

(*a*) USWB open pan evaporimeter

(*b*) Sunken screen evaporimeter

(*c*) Can evaporimeter

(*d*) All

Q. 102 The equipment not used for measurement of evaporation:

(*a*) USWB open pan evaporimeter

(*b*) Sunscreen evaporimeter

(*c*) Can evaporimeter

(*d*) Portable evaporimeter

Q. 103 The height of cylinder in USWB open pan evaporimeter is:

(*a*) 25 cm (*b*) 250cm

(*c*) 122 cm (*d*) 122 mm

Q. 104 The diameter of cylinder in USWB open pan evaporimeter is:

(*a*) 25 cm (*b*) 250cm

(*c*) 122 cm (*d*) 122 mm

Q. 105 Evaporation is measured in:

(*a*) mm/day (*b*) cm/day

(*c*) kg/day (*d*) cm/min

Q. 106 The height of cylindrical vessel in sunken screen evaporimeter is:

(*a*) 4.5 cm (*b*) 45cm

(*c*) 450 cm (*d*) 122 mm

Q. 107 The diameter of cylindrical vessel in sunken screen evaporimeter is:

(*a*) 4.5 cm (*b*) 45cm

(*c*) 60 cm (*d*) 60 mm ans-c

Q. 108 Equipment used to measure evapotranspiration under field condition is:

(*a*) USWB open pan evaporimeter

(*b*) Sunken screen evaporimeter

(*c*) Can evaporimeter

(*d*) Both (*b*) and (*c*)

Q. 109 Portable evaporimeter was developed in:

(*a*) India (*b*) Indenosia

(*c*) Israel (*d*) USA

Q. 110 Portable evaporimeter consists of:

(*a*) Glass tube (*b*) Iron tube

(*c*) Plastic tube (*d*) Aluminium tube

Q. 111 The falling down of condensed moisture to ground surface is called:

(*a*) Mist (*b*) Precipitation

(*c*) Fog (*d*) Hail storm

Q. 112 When the water droplets evaporate before it reaches the ground it is called:

(*a*) Mist (*b*) Precipitation

(*c*) Fog (*d*) Hail storm

Q. 113 When rain falls on the ground and forms a sheet or coating of ice it is called:

(*a*) Mist (*b*) Glaze

(*c*) Fog (*d*) Hail storm

Q. 114 Rime is a:

(*a*) Fog (*b*) Glaze

(*c*) Freezing fog (*d*) Hail storm

Q. 115 The solid precipitation on the ground in the form of ice crystals it is called:

(*a*) Snow (*b*) Glaze

(*c*) Freezing fog (*d*) Hail storm

Q. 116 The solid precipitation on the ground in the form of small particles of clear ice it is called:

(*a*) Snow (*b*) Glaze

(*c*) Sleet (*d*) Hail storm

Q. 117 The solid precipitation in the form of large, hard rounded pallets of ice and compact snow it is called:

(*a*) Snow (*b*) Glaze

(*c*) Sleet (*d*) Hail storm

Q. 118 Instrument used for measurement of rainfall continuously:

(*a*) Rain guage

(*b*) Automatic rain guage

(*c*) Screw guage

(*d*) None

Q. 119 The aggregate of minute drops of water suspended in air at higher altitude is called:

(*a*) Snow (*b*) Glaze

(*c*) Sleet (*d*) Cloud

Q. 120 Which of the following is not a basic form of cloud:

(*a*) Cirrus (*b*) Stratus

(*c*) Cumulus (*d*) Cumulo-nimbus

Q. 121 In the clouds associated with rains, the word prefixed or suffixed to the basic form is:

(*a*) Alto (*b*) Nimbus

(*c*) Cumulus (*d*) Cumulo-nimbus

Q. 122 The clouds are classified in to .... characteristic forms:

(*a*) 3 (*b*) 5

(*c*) 7 (*d*) 10

Q. 123 The highest clouds are:

(*a*) Cirrus (*b*) Stratus

(*c*) Cumulus (*d*) Cumulo-nimbus

Q. 124 The clouds associated with rains in middle and high altitude is:

(*a*) Cirrus (*b*) Stratus

(*c*) Alto- cumulus (*d*) Cumulo-nimbus

Q. 125 The cloud in the form of uniform grey cloud sheet:

(*a*) Cirrus (*b*) Stratus

(*c*) Cumulus (*d*) Cumulo-nimbus

Q. 126 The clouds associated with steady precipitation:

(*a*) Cirrus (*b*) Nimbo-Stratus

(*c*) Cumulus (*d*) Cumulo-nimbus

Q. 127 Clouds prominent during summer:

(*a*) Cirrus (*b*) Stratus

(*c*) Cumulus (*d*) Cumulo-nimbus

Q. 128 Clouds associated with thunder storms:

(*a*) Cirrus (*b*) Stratus

(*c*) Cumulus (*d*) Cumulo-nimbus

Q. 129 The outer solid crust of earth is called:

(*a*) Atmosphere (*b*) Lithosphere

(*c*) Hydrosphere (*d*) None

Q. 130 The total volume of water contained in oceans is about:

(*a*) 1.94×10^9 Km^3 (*b*) 1.64×10^9 Km^3

(*c*) 1.96×10^9 Km^3 (*d*) 1.46×10^9 Km^3

Q. 131 The average annual world precipitation is:

(*a*) 1500mm (*b*) 1200mm

(*c*) 1000mm (*d*) None

Q. 132 Temperature is high near the:

(*a*) Poles (*b*) Equator

(*c*) Remains unchanged (*d*) None

Q. 133 The low pressure areas near the equator is called:

(*a*) Oldrums (*b*) Doldrums

(*c*) Polar calms (*d*) None

Q. 134 Horse latitude is the belt between:

(*a*) 30 to 35° North latitude

(*b*) 30 to 35° South latitude

(*c*) Both (*a*) and (*b*)

(*d*) None

Q. 135 The decrease in air pressure due to movement of air from polar region is known as:

(*a*) Polar storms (*b*) Doldrums

(*c*) Polar calms (*d*) None

Q. 136 The normal date of arrival of monsoon in India is:

(*a*) 1st June (*b*) 15th June

(*c*) 20th June (*d*) 30th June

Q. 137 The grand period of rainfall in India is:

(*a*) South-west monsoon (*b*) South-east monsoon

(*c*) North-east monsoon (*d*) None

Q. 138 The part of India not receiving rains from south-west monsoon is:

(*a*) South-west Tamilnadu

(*b*) East coast of Tamilnadu

(*c*) Kerala

(*d*) None

Q. 139 South coastal Andhra Pradesh and Tamilnadu receives rainfall from:

(*a*) South-west monsoon (*b*) South-east monsoon

(*c*) North-east monsoon (*d*) None

Q. 140 The short range weather forecast is for a period of:

(*a*) 1–2 days (*b*) 3–4 days to 2 weeks

(*c*) More than four weeks (*d*) Less than four weeks

Q. 141 The medium range weather forecast is for a period of:

(*a*) 1–2 days (*b*) 3–4 days to 2 weeks

(*c*) More than four weeks (*d*) Less than four weeks

Q. 142 The long range weather forecast is for a period of:

(*a*) 1–2 days (*b*) 3–4 days to 2 weeks

(*c*) More than four weeks (*d*) Less than four weeks

Q. 143 Which of the following is not a method of weatner forecast:

(*a*) Synoptic method (*b*) Statistical method

(*c*) Numerical method (*d*) Asynoptic method

Q. 144 Cloud seeding is done with the help of:

(*a*) Silver iodide (*b*) Magnesium iodide

(*c*) Strontium chloride (*d*) All

Q. 145 The technique of cloud seeding aims at correcting the:

(*a*) Deficiencies of nuclei

(*b*) Excess of nuclei

(*c*) Deficiencies of neutrons

(*d*) All

Q. 146 Silver iodide is used for:

(*a*) Warm cloud seeding (*b*) Cold cloud seeding

(*c*) Both (*d*) None

Q. 147 Sodium chloride is used for:

(*a*) warm cloud seeding (*b*) Cold cloud seeding

(*c*) Both (*d*) None

Q. 148 The adverse effect of high wind velocity can be reduced by:

(*a*) Wind breaks (*b*) Shelterbelts

(*c*) Both (*d*) None

Q. 149 India has been divided into:

(*a*) 12 agro climatic zones

(*b*) 15 agro climatic zones

(*c*) 18 agro climatic zones

(*d*) None

Q. 150 Radiation harmful to the plants comes in the spectral region:

(*a*) <0.4μm (*b*) 0.4μm–0.7μm

(*c*) 0.7μm–1.0μm (*d*) > 0.4μm

Q. 151 Photosynthetically active radiation comes in the spectral region:

(*a*) < 0.4 µm (*b*) 0.4 µm–0.7 µm

(*c*) 0.7 µm–1.0 µm (*d*) > 0.4 µm

Q. 152 Near infrared radiation comes in the spectral region:

(*a*) < 0.4 µm (*b*) 0.4 µm–0.7 µm

(*c*) 0.7 µm–3.0 µm (*d*) > 0.4 µm

Q. 153 Radiation having influence on seed germination and dormancy of seeds:

(*a*) Near infrared

(*b*) Photosynthetically active radiation

(*c*) Ultraviolet

(*d*) None

Q. 154 Chlorophyll formation is promoted by light in the spectral region:

(*a*) < 0.25µm (*b*) 0.3µm–0.338µm

(*c*) 0.4µm–1.0µm (*d*) > 1.0µm

Q. 155 Phytochrome is excited by:

(*a*) Green light (*b*) Red light

(*c*) White light (*d*) All

Q. 156 The photosynthetic electron transport chain is formed during subsequent development of chloroplast under:

(*a*) Green light (*b*) Red light

(*c*) White light (*d*) All

Q. 157 Formation of protein-pigment system of the chloroplast and their resistance to senescence is due to:

(*a*) Green light (*b*) Red light

(*c*) White light (*d*) All

Q. 158 The amount of solar radiation present in any layer of the crop can be calculated by:

(*a*) $I = I_0 e^{-KLAI}$ (*b*) $I = I_0 e^{-K}$

(*c*) $I_0 = I e^{-KLAI}$ (*d*) $I = I_0 e^{-LAI}$

Q. 159 The effect of light on plants is known as:

(*a*) Thermoperiodism (*b*) Photoperiodism

(*c*) Both (*d*) None

Q. 160 Long day plants flower when the length of light is:

(*a*) <14 hrs (*b*) >14 hrs

(*c*) <10 hrs (*d*) >10 hrs ans-b

Q. 161 Short day plants flower when the length of light is:

(*a*) < 14 hrs (*b*) > 14 hrs

(*c*) < 10 hrs (*d*) > 10 hrs

Q. 162 Cockelbur is an example of:

(*a*) Qualitative short day plant

(*b*) Quantitative short day plant

(*c*) Qualitative long day plant

(*d*) Quantitative short day plant

Q. 163 At the time of equinoxes:

(*a*) The length of day and night is equal

(*b*) The length of day and night is not equal

(*c*) The length of light is equal

(*d*) None

Q. 164 The rate of increase in reaction for every 10°C in temperature is called:

(*a*) Q_{10} (*b*) K_{10}

(*c*) C_{10} (*d*) T_{10} ans-a:

Q. 165 The relationship between growth duration and temperature is termed as:

(*a*) Base temperature (*b*) Growing degree days

(*c*) Photothermal unit (*d*) None

Q. 166 Growing degree days can mathematically be expressed as:

(*a*) $\sum_{i=1}^{n} \left(\frac{T_{max} + T_{min}}{2} + T_t \right)$

(b) $\sum_{i=1}^{n}\left(T_t - \frac{T_{max}+T_{min}}{2}\right)$

(c) $\sum_{i=1}^{n}\left(\frac{T_{max}+T_{min}}{2} - T_t\right)$

(d) $\sum_{i=1}^{n}\left[T_t - \left(T_{max}+T_{min}\right)\right]$

Q. 167 The lowest temperature below which there is no growth is called as:

(a) Q_{10} (b) Base temperature

(c) Photothermal unit (d) T_{10}

Q. 168 Base temperature for wheat is:

(a) 4.5°C (b) 45°C

(c) 10°C (d) 1.0°C

Q. 169 Base temperature for rice is:

(a) 4.5°C (b) 45°C

(c) 10°C (d) 1.0°C

Q. 170 Coefficient of variation is given by:

(a) SD/Mean × 100 (b) 45°C

(c) 1.0°C (d) 1.0°C

Q. 171 Q_{10} can be calculated by:

(a) $Q10 = \frac{\text{Rate of reaction at (t+10)°C}}{\text{Rate of reaction at t°C}}$

(b) $Q10 = \frac{\text{Rate of reaction at t°C}}{\text{Rate of reaction at (t+10)°C}}$

(c) $Q10 = \frac{\text{Rate of reaction at (t–10)°C}}{\text{Rate of reaction at t°C}}$

(d) $Q10 = \frac{\text{Rate of reaction at t°C}}{\text{Rate of reaction at (t–10)°C}}$

ANSWERS

1	(*b*)	23	(*a*)	45	(*a*)	67	(*c*)
2	(*b*)	24	(*b*)	46	(*b*)	68	(*b*)
3	(*a*)	25	(*c*)	47	(*c*)	69	(*b*)
4	(*b*)	26	(*d*)	48	(*a*)	70	(*b*)
5	(*a*)	27	(*b*)	49	(*b*)	71	(*b*)
6	(*c*)	28	(*b*)	50	(*c*)	72	(*b*)
7	(*b*)	29	(*a*)	51	(*c*)	73	(*b*)
8	(*c*)	30	(*b*)	52	(*a*)	74	(*a*)
9	(*d*)	31	(*c*)	53	(*b*)	75	(*b*)
10	(*d*)	32	(*b*)	54	(*a*)	76	(*a*)
11	(*c*)	33	(*a*)	55	(*b*)	77	(*b*)
12	(*a*)	34	(*b*)	56	(*b*)	78	(*c*)
13	(*d*)	35	(*a*)	57	(*b*)	79	(*b*)
14	(*b*)	36	(*b*)	58	(*b*)	80	(*b*)
15	(*d*)	37	(*b*)	59	(*a*)	81	(*b*)
16	(*a*)	38	(*c*)	60	(*b*)	82	(*b*)
17	(*a*)	39	(*c*)	61	(*b*)	83	(*d*)
18	(*d*)	40	(*b*)	62	(*d*)	84	(*a*)
19	(*b*)	41	(*b*)	63	(*d*)	85	(*b*)
20	(*c*)	42	(*c*)	64	(*c*)	86	(*c*)
21	(*a*)	43	(*b*)	65	(*a*)	87	(*a*)
22	(*b*)	44	(*d*)	66	(*a*)	88	(*b*)

89	(*c*)	110	(*b*)	131	(*c*)	152	(*c*)
90	(*c*)	111	(*b*)	132	(*b*)	153	(*a*)
91	(*c*)	112	(*a*)	133	(*b*)	154	(*b*)
92	(*a*)	113	(*b*)	134	(*c*)	155	(*a*)
93	(*d*)	114	(*c*)	135	(*c*)	156	(*c*)
94	(*c*)	115	(*a*)	136	(*a*)	157	(*b*)
95	(*c*)	116	(*c*)	137	(*a*)	158	(*a*)
96	(*a*)	117	(*d*)	138	(*b*)	159	(*b*)
97	(*b*)	118	(*b*)	139	(*c*)	160	(*b*)
98	(*c*)	119	(*d*)	140	(*a*)	161	(*c*)
99	(*c*)	120	(*d*)	141	(*b*)	162	(*a*)
100	(*a*)	121	(*b*)	142	(*c*)	163	(*a*)
101	(*d*)	122	(*d*)	143	(*d*)	164	(*a*)
102	(*b*)	123	(*a*)	144	(*a*)	165	(*b*)
103	(*a*)	124	(*c*)	145	(*a*)	166	(*c*)
104	(*c*)	125	(*b*)	146	(*b*)	167	(*b*)
105	(*a*)	126	(*b*)	147	(*a*)	168	(*a*)
106	(*b*)	127	(*c*)	148	(*c*)	169	(*c*)
107	(*c*)	128	(*d*)	149	(*b*)	170	(*a*)
108	(*d*)	129	(*b*)	150	(*a*)	171	(*a*)
109	(*c*)	130	(*d*)	151	(*b*)		

Chapter 5
Remote Sensing

Q. 1 Which of the following was the first Indian experimental communication satellite:

(*a*) Aryabhat (*b*) Apple

(*c*) IRS-1A (*d*) Resourcesat

Q. 2 Which of the following sensors was/were used in IRS-1C:

(*a*) PAN (*b*) LISS

(*c*) WiFS (*d*) All

Q. 3 Which of the following satellite is the most recent one:

(*a*) IRS-1D (*b*) IRS-P2

(*c*) Oceansat-1 (*d*) Risat

Q. 4 Which of the following band of LISS camera of IRS-1A/B satellite is used for chlorophyll absorption:

(*a*) Band 1 (*b*) Band 2

(*c*) Band 3 (*d*) Band 1 and 3

Q. 5 Spatial resolution of Cartosat-2 satellite is:

(*a*) nearly 1 m (*b*) 5.8 m

(*c*) 188 m (*d*) 32 m

Q. 6 Which of the following band is categorized for visible wavelength in Multi Spectral Scanner:

(*a*) 1 (*b*) 2

(*c*) 3 (*d*) All

Q. 7 What is the distance of sun-synchronous satellite from the earth surface:

(*a*) 600–1000 km (*b*) 1000–2000 km

(*c*) 2000–3000 km (*d*) 36000 km

Q. 8 Which of the following resolution describes (the capability of identifying nearly kept two object):

(*a*) Spatial (*b*) Temporal

(*c*) Radiometric (*d*) Spectral

Q. 9 NDVI stand for:

(*a*) Normalized Difference Vegetation Index

(*b*) Nadir View

(*c*) Net Drought Vegetation Index

(*d*) National Drought and Vegetation Institute

Q. 10 NDVI can be estimated by the manipulation of which of the following band:

(*a*) Red (*b*) Near Infrared

(*c*) Both (*d*) None

Q. 11 If spatial resolution is smaller, what will be the resolving power of the sensor:

(*a*) Greater (*b*) Smaller

(*c*) Same (*d*) None

Q. 12 Which of the following resolution best describe (sensitivity of the sensor to incoming radiane):

(*a*) Spatial (*b*) Temporal

(*c*) Radiometric (*d*) Spectral

Q. 13 Which of the following resolution best describe "the acquisition data has been recorded on 4 dates":

(*a*) Spatial (*b*) Temporal

(*c*) Radiometric (*d*) Spectral

Q. 14 The minimum distance between two objects that can be distinguish by continuous different wavelength range is referred as:

(*a*) Spatial (*b*) Temporal

(*c*) Radiometric (*d*) Spectral

Q. 15 Quantitative measurement of the properties of an object at one or several wavelength interval is called as:

(*a*) Spectral signature (*b*) Spectral reflectance

(*c*) Spectral region (*d*) Spectral response

Q. 16 Microwave band is more useful in case of:

(*a*) Cloud (*b*) Dust

(*c*) Haze (*d*) All

Q. 17 Which of the following radiation is used in passive remote sensor:

(*a*) Sun (*b*) Vapour light

(*c*) Tube light (*d*) All

Q. 18 IRS- 1C/1D PAN spatial resolution is:

(*a*) 5.8 m (*b*) 24 m

(*c*) 188 m (*d*) 30 m

Q. 19 PAN camera of IRS 1C/1D provide which of the following type of image:

(*a*) Coloured (*b*) Blue and red

(*c*) Blue and green (*d*) Black and white

Q. 20 Repeatitivity of IRS- 1C/1D satellite is:

(*a*) 16 days (*b*) 24 days

(*c*) 3 days (*d*) 1 hour

Q. 21 In which of the following condition reflectance % will be maximum:

(*a*) Dry soil (*b*) Wet soil

(*c*) ploughed soil (*d*) Water bodies

Q. 22 What could be the reflectance % if the farmer is applying FYM in their field:

(*a*) Increased

(*b*) Decreased

(*c*) No effect

(*d*) First increases then decreases

Q. 23 The pixels of the band in the sensor which is converted into digital form is referred as:

(*a*) Digital band (*b*) Digital sensor

(*c*) Digital map (*d*) Digital number

Q. 24 Satellite image showing red colour indicates:

(*a*) Fire (*b*) Forest

(*c*) Ocean (*d*) Ice cap

Q. 25 Satellite image showing white colour indicates:

(*a*) Sand (*b*) Ice cap

(*c*) Saline soil (*d*) All

Q. 26 Which of the following kind of wave is related to atmospheric window:

(*a*) Long wave (*b*) Short wave

(*c*) Both (*d*) None

Q. 27 Remote sensing satellite is known as:

(*a*) polar satellite

(*b*) Sun-synchronous satellite

(*c*) Both

(*d*) None

Q. 28 Which of the following wavelength is normally used in remote sensing:

(*a*) Visible (*b*) Near infrared

(*c*) Both (*d*) None

Q. 29 Which of the following band is used in infrared:

(*a*) 0.4–0.7 nm (*b*) 0.7–1.3 nm

(*c*) 1.3–3.0 nm (*d*) 3.0–100 nm

Q. 30 Ratio Vegetation Index can be calculated by:

(*a*) Red/Near Infrared (*b*) Near Infrared/Red

(*c*) Near Infrared-Red (*d*) Red-Infrared

Q. 31 LISS stands for:

(*a*) Linear Imaging Self Scanner

(*b*) Lime Imaging Scanning System

(*c*) Length Imaging Scanning System

(*d*) None

Q. 32 WiFS stands for:

(*a*) Wide Field Scanner (*b*) Wield Field Scanner

(*c*) Both (*d*) None

Q. 33 Essentially remote sensing has how many component:

(*a*) Signal (*b*) Sensor

(*c*) Sensing (*d*) All

Q. 34 FWHM stands for:

(*a*) Full Width at Half Maxima

(*b*) Full and Wide Hazy Monitoring

(*c*) Fertilizer Widely and Hormones Monitoring

(*d*) None

Q. 35 Which of the following is a type of signal used in remote sensing:

(*a*) Acaustic signal (*b*) Particulate signal

(*c*) Electromagnetic signal (*d*) All

Q. 36 Which of the following is medium resolution satellite:

(*a*) SPOT 1-3 (*b*) CBERS

(*c*) IRS 1C/1D (*d*) All

Q. 37 Which of the following is high resolution satellite:

(*a*) COSMOS (*b*) EROS

(*c*) ICONOS (*d*) All

Q. 38 Which of the following medium resolution satellite has highest number of bands:

(*a*) LANDSAT-7 (*b*) IRS 1C/1D

(*c*) IRS 1A (*d*) Apple

Q. 39 Cartosat satellites is largely meant for:

(*a*) Mapping

(*b*) Disaster management

(*c*) Information Technology

(*d*) Atmospheric information

Q. 40 Which of the following spectraı rays is used for mapping of radioactive material:

(*a*) Gamma (*b*) UV

(*c*) Visible (*d*) Infrared

Q. 41 Which of the following spectral rays is used for presence of hydrogen and helium in atmosphere:

(*a*) Gamma (*b*) UV

(*c*) Visible (*d*) Infrared

Q. 42 Which of the following spectral rays is used for vegetation properties:

(*a*) Gamma (*b*) UV

(*c*) Visible (*d*) None

Q. 43 The infrared thermometer contain which of the following component:

(*a*) Filter (*b*) Detector

(*c*) Read out (*d*) all

Q. 44 A range of wavelength of electromagnetic radiation is known as:

(*a*) Band (*b*) Sensor

(*c*) Plateform (*d*) Resolution

Q. 45 Alters data to correspond with true ground or image space in a known coordinate system is known as:

(*a*) Geometric correction (*b*) Geodatic correction

(*c*) Thematic correction (*d*) Spatial correction

Q. 46 A raster module in the Arc/infoGIS package is known as:

(*a*) Pixel (*b*) Grid

(*c*) Digital number (*d*) None

Q. 47 To verify the correctness of remote sensing data by field studies is known as:

(*a*) False colour composit (*b*) Grid

(*c*) Ground truth (*d*) Pixel value

Q. 48 An interpolation technique for obtaining statistically unbiased estimates of surface elevations from a set of control points

(*a*) Interpolation (*b*) Krigging

(*c*) Label (*d*) Precision

Q. 49 The part of the drawn map explaining the meaning of symbols used to code the depicted geographical elements is known as:

(*a*) Layer (*b*) Label

(*c*) Legend (*d*) None

Q. 50 A two-dimentional graphic representation of the earth's surface that displays spatial relationships among features:

(*a*) Grid (*b*) Interpolation

(*c*) Map (*d*) Plateform

Q. 51 Irregular variation, usually short range that cannot be easily explained or associated with major mapped features or processes is referred as:

(*a*) Krigging (*b*) Noise

(*c*) Output (*d*) Theme

Q. 52 Quantum detectors are of how many types:

(*a*) Photoemmisive

(*b*) Photoconductive

(*c*) Photovoltaic

(*d*) All

Q. 53 Natural gamme rays has a wavelength of:

(*a*) 30 pm (*b*) 0.03 pm

(*c*) 0.00003 pm (*d*) None

Q. 54 Artificial gamme rays has a wavelength of:

(*a*) 30 pm (*b*) 0.03 pm

(*c*) 0.00003 pm (*d*) None

Q. 55 Cosmic gamme rays has a wavelength of:

(*a*) 30 pm (*b*) 0.03 pm

(*c*) 0.00003 pm (*d*) None

Q. 56 An increase in soil moisture will result in rapid:

(*a*) Increase in reflectence

(*b*) Decrease in reflectence

(*c*) No effect in reflectence

(*d*) None

Q. 57 An increase in sand content will result in rapid:

(*a*) Increase in reflectence

(*b*) Decrease in reflectence

(*c*) No effect in reflectence

(*d*) None

Q. 58 An increase in types of clay will result in rapid:

(*a*) Increase in reflectence

(*b*) Decrease in reflectence

(*c*) No effect in reflectence

(*d*) None

Q. 59 An increase in Lime stone will result in rapid:

(*a*) Increase in reflectence

(*b*) Decrease in reflectence

(*c*) No effect in reflectence

(*d*) None

Q. 60 An increase in igneous rock will result in rapid:

(*a*) Increase in reflectence

(*b*) Decrease in reflectence

(*c*) No effect in reflectence

(*d*) None

Q. 61 An increase in organic matter will result in rapid:

(*a*) Increase in reflectence

(*b*) Decrease in reflectence

(*c*) No effect in reflectence

(*d*) None

Q. 62 An increase in iron oxide will result in rapid:

(*a*) Increase in reflectence

(*b*) Decrease in reflectence

(*c*) No effect in reflectence

(*d*) None

Q. 63 An increase in salinity will result in rapid:

(*a*) Increase in reflectence

(*b*) Decrease in reflectence

(*c*) No effect in reflectence

(*d*) None

Q. 64 An increase in soil colour will result in rapid:

(*a*) Increase in reflectence

(*b*) Decrease in reflectence

(*c*) No effect in reflectence

(*d*) None

Q. 65 The ratio which is a sensitive indicator of green biomass is given by:

(*a*) NIR/IR (*b*) NDVI

(*c*) RVI (*d*) None

Q. 66 Maximum absorption of chlorophyll *a* is denoted by which of the following wavelength range:

(*a*) 430 and 662 nm (*b*) 450 and 470 nm

(*c*) 480 and 540 nm (*d*) 610 and 640 nm

Q. 67 Maximum absorption of carotenoids is denoted by which of the following wavelength range:

(*a*) 430 and 662 nm (*b*) 450 and 470 nm

(*c*) 480 and 540 nm (*d*) 610 and 640 nm

Q. 68 Maximum absorption of xyanthophyll is denoted by which of the following wavelength range:

(*a*) 430 and 662 nm (*b*) 450 and 470 nm

(*c*) 480 and 540 nm (*d*) 610 and 640 nm

Q. 69 Maximum absorption of phycocyanin is denoted by which of the following wavelength range:

(*a*) 430 and 662 nm (*b*) 450 and 470 nm

(*c*) 480 and 540 nm (*d*) 610 and 640 nm

Q. 70 Passive microwave energy can be emmited by the atmosphere by which of the following way

(*a*) Reflection from the surface

(*b*) Emission from the surface

(*c*) Transmission from the surface

(*d*) All

Q. 71 Passive microwave sensors are characterised by:

(*a*) Low spatial resolution

(*b*) High spatial resolution

(*c*) Both

(*d*) None

Q. 72 Application of passive microwave remote sensi g includes:

(*a*) Meteorology (*b*) Hydrology

(*c*) Oceanography (*d*) All

Q. 73 Target interaction of microwave remote sensing is affected by which of the following:

(*a*) Surface roughness (*b*) Radar viewing

(*c*) moisture content (*d*) All

Q. 74 The image data which is acquired from remote sensing system are stored in which kind of format:

(*a*) Band sequential

(*b*) Band interleaved by lime

(*c*) Band interleaved by lime

(*d*) All

Q. 75 The point on the ground vertically beneath the respective centre of the camera lens is known as:

(*a*) NADIR (*b*) Zenith

(*c*) Both (*d*) None

Q. 76 The smallest discrete element which make up a raster image is referred as:

(*a*) Digital number (*b*) pixel

(*c*) Noise (*d*) Point

Q. 77 To find specific information in a database is referred as:

(*a*) Projection (*b*) Node

(*c*) Query (*d*) Registration

Q. 78 The scanned area of cathode ray tube is referred as:

(*a*) Raster (*b*) Vector

(*b*) Both (*c*) None

Q. 79 Remote sensing is aquiring knowledge from a distant place by what actual touch with object:

(*a*) With (*b*) Without

(*c*) Both (*d*) None

Q. 80 Manipulations and functions that operate on raster object on cell by cell is:

(*a*) Raster data (*b*) Raster algebra

(*c*) Query (*d*) Point

Q. 81 An Interpollating polynomial for a set of coordinate points used to fit a curve that connects the points is:

(*a*) Spline (*b*) Slope

(*c*) Scale (*d*) Theme

Q. 82 The ratio or fraction between the size of an object on a man and its size in the real world is:

(*a*) Map (*b*) Resolution

(*c*) Slope (*d*) Scale

Q. 83 Processes to convert analogue information from maps, photographs or overlays into a digital format usable by a computer is:

(*a*) Scanning (*b*) Resolution

(*c*) Topple (*d*) Topography

Q. 84 A map related to a single topic theme or subject of discourse is also called as:

(*a*) Thematic map (*b*) Contour map

(*c*) Soil map (*d*) None

Q. 85 A reference to the values of surface height in a digital elevation/terrain model at a particular X,Y location is referred as:

(*a*) X-value (*b*) Y-value

(*c*) Z-value (*d*) All

Q. 86 GIS function in cartography is to:

(*a*) Data acquisition (*b*) Mapping

(*c*) Display (*d*) All

Q. 87 GIS function in remote sensing is to:

(*a*) Data acquisition (*b*) Mapping

(*c*) Display (*d*) All

Q. 88 GIS function in statistics is to:

(a) Data base (b) Mapping

(c) Display (d) All

Q. 89 GIS function in survey is to:

(a) Data acquisition (b) Mapping

(c) Both (d) None

Q. 90 GIS function in Geography is to:

(a) Spatial analysis (b) Mapping

(c) Display (d) All

Q. 91 Who was the first person who presented a world atlas:

(a) Ptolemy (b) I Newton

(c) Gerhard Mercator (d) A. Penck

Q. 92 Who was the first person who presented a Globe:

(a) Ptolemy (b) I Newton

(c) Gerhard Mercator (d) A. Penck

Q. 93 Who was the first person who presented that the earth is flattened at the poles:

(a) Ptolemy (b) I Newton

(c) Gerhard Mercator (d) A. Penck

Q. 94 Point data are depicted on images with:

(a) Graphical science (b) Continuous line

(c) Both (d) None

Q. 95 Line data are depicted on images with:

(a) Graphical science (b) Continuous line

(c) Both (d) None

Q. 96 Position of view point on a map projection system can be a:

(a) Gnomonic (b) Stereographic

(c) Orthographic (d) All

Q. 97 Position of a tangent surface on a map projection system can be a:

(*a*) polar (*b*) Eque distant

(*c*) Oblique (*d*) All

Q. 98 The most common raster data format are:

(*a*) Tiff (*b*) Gif

(*c*) JPEG (*d*) All

Q. 99 Precision is more in:

(*a*) Vector data structure (*b*) Raster data structure

(*c*) Both (*d*) None

Q. 100 Which of the following data structure can best describe topology:

(*a*) Vector data structure (*b*) Raster data structure

(*c*) Both (*d*) None

Q. 101 Which of the following data structure can best describe update:

(*a*) Vector data structure (*b*) Raster data structure

(*c*) Both (*d*) None

Q. 102 Which of the following data structure can best describe coordinate transformation:

(*a*) Vector data structure (*b*) Raster data structure

(*c*) Both (*d*) None

Q. 103 Which of the following data structure can best describe simulation modelling:

(*a*) Vector data structure (*b*) Raster data structure

(*c*) Both (*d*) None

Q. 104 Which of the following data structure can best describe graphic update:

(*a*) Vector data structure (*b*) Raster data structure

(*c*) Both (*d*) None

Q. 105 Which of the following data structure can best describe data structure:

(*a*) Vector data structure (*b*) Raster data structure

(*c*) Both (*d*) None

Q. 106 Which of the following equipment is the best choice for analog map kind of data source:

(*a*) Digitizer (*b*) Stereo plotter

(*c*) Image zoom scope (*d*) GPS

Q. 107 Which of the following equipment is the best choice for aerial photographs kind of data source:

(*a*) Digitizer (*b*) Stereo plotter

(*c*) Image zoom scope (*d*) GPS

Q. 108 Which of the following equipment is the best choice for satellite imagery kind of data source:

(*a*) Digitizer (*b*) Stereo plotter

(*c*) Image zoom scope (*d*) GPS

Q. 109 Which of the following equipment is the best choice for ground survey kind of data source:

(*a*) Digitizer (*b*) Stereo plotter

(*c*) Image zoom scope (*d*) GPS

Q. 110 Which of the following scanner is having very high resolution:

(*a*) Mechanical scanner (*b*) Video camera

(*c*) CCD camera (*d*) None

Q. 111 Which of the following scanner is having low resolution:

(*a*) Mechanical scanner (*b*) Video camera

(*c*) CCD camera (*d*) None

Q. 112 Which of the following scanner is having medium resolution:

(*a*) Mechanical scanner (*b*) Video camera

(*c*) CCD camera (*d*) None

ANSWERS

1	(*b*)	23	(*d*)	45	(*a*)	67	(*b*)
2	(*d*)	24	(*b*)	46	(*b*)	68	(*c*)
3	(*d*)	25	(*d*)	47	(*c*)	69	(*d*)
4	(*d*)	26	(*a*)	48	(*b*)	70	(*d*)
5	(*a*)	27	(*c*)	49	(*c*)	71	(*a*)
6	(*d*)	28	(*c*)	50	(*c*)	72	(*d*)
7	(*a*)	29	(*b*)	51	(*b*)	73	(*d*)
8	(*a*)	30	(*b*)	52	(*d*)	74	(*d*)
9	(*a*)	31	(*a*)	53	(*a*)	75	(*a*)
10	(*c*)	32	(*a*)	54	(*b*)	76	(*b*)
11	(*a*)	33	(*d*)	55	(*c*)	77	(*c*)
12	(*c*)	34	(*a*)	56	(*b*)	78	(*a*)
13	(*b*)	35	(*d*)	57	(*a*)	79	(*b*)
14	(*d*)	36	(*d*)	58	(*b*)	80	(*b*)
15	(*a*)	37	(*d*)	59	(*a*)	81	(*a*)
16	(*d*)	38	(*a*)	60	(*b*)	82	(*d*)
17	(*d*)	39	(*a*)	61	(*b*)	83	(*a*)
18	(*a*)	40	(*a*)	62	(*b*)	84	(*a*)
19	(*d*)	41	(*b*)	63	(*a*)	85	(*c*)
20	(*b*)	42	(*c*)	64	(*c*)	86	(*d*)
21	(*a*)	43	(*d*)	65	(*c*)	87	(*d*)
22	(*b*)	44	(*a*)	66	(*a*)	88	(*d*)

89	(*c*)	95	(*b*)	101	(*b*)	107	(*b*)
90	(*d*)	96	(*d*)	102	(*b*)	108	(*c*)
91	(*a*)	97	(*d*)	103	(*a*)	109	(*d*)
92	(*c*)	98	(*d*)	104	(*b*)	110	(*a*)
93	(*b*)	99	(*a*)	105	(*a*)	111	(*b*)
94	(*a*)	100	(*a*)	106	(*a*)	112	(*c*)

Chapter 6
GPS and GIS

GPS

Q. 1 GPS stand for:

(*a*) Global Positioning System
(*b*) Geographical Positioning System
(*c*) Geological Positioning System
(*d*) Geo Positioning System

Q. 2 GPS provide:

(*a*) Latitude (*b*) Longitude
(*c*) Altitude (*d*) All of the above

Q. 3 GPS provides:

(*a*) Latitude (*b*) Area
(*c*) Distance (*d*) All

Q. 4 GPS works on the principles of:

(*a*) X value (*b*) Y value
(*c*) Z value (*d*) All

Q. 5 Total number of satellites in a GPS system are:

(*a*) 18 (*b*) 21
(*c*) 23 (*d*) 24

Q. 6 Total number of satellites in use ina GPS system is:

(*a*) 18 (*b*) 21
(*c*) 23 (*d*) 24

Q. 7 Number of satellites spares in a GPS system is:

(*a*) 3 (*b*) 6
(*c*) 9 (*d*) 12

Q. 8 Which GPS system is available:

(*a*) NAVSTAR (*b*) GLONAS

(*c*) Both (*d*) None

Q. 9 Accuracy of GPS:

(*a*) Vary (*b*) Accurate

(*c*) Both (*d*) None

Q. 10 If a person is lost in Jhum Forest, can he come out with the help of a GPS:

(*a*) Yes (*b*) No

(*c*) Never (*d*) Cannot say

Q.11 If a person wants to make a map with the help of a GPS Can he do it:

(*a*) Yes (*b*) No

(*c*) Never (*d*) Cannot say

Q. 12 GPS is weather:

(*a*) Dependent (*b*) Independent

(*c*) Both (*d*) None

Q. 13 Components of a GPS is/are:

(*a*) Space segment (*b*) Control segment

(*c*) User segment (*d*) All

Q. 14 Number of Orbits available in a GPS system is:

(*a*) 2 (*b*) 3

(*c*) 4 (*d*) 6

Q. 15 How many satellites are existing in one orbit of a GPS system:

(*a*) 2 (*b*) 4

(*c*) 6 (*d*) 8

Q. 16 GPS satellites are fixed at a distance of:

(*a*) 20,200km (*b*) 10,200km

(*c*) 900km (*d*) 600km

Q. 17 The orbital period of each GPS satellite is:

(*a*) 6 hrs (*b*) 4hrs

(*c*) 12hrs (*d*) 10hrs

Q. 18 Each orbital plane of satellite in FPS is inclined at - equatorial plane of the earth:

(*a*) 55 degree (*b*) 45 degree

(*c*) 35 degree (*d*) 60 degree

Q. 19 The orbital plane of the satellite is displaced by which of the following longitude:

(*a*) 100 degree (*b*) 110 degree

(*c*) 120 degree (*d*) 90 degree

Q. 20 NAVSTAR was launched in the year:

(*a*) 1978 (*b*) 1968

(*c*) 1958 (*d*) 1982

Q. 21 The problem in NAVSTAR was rectified in the year:

(*a*) 1994 (*b*) 1984

(*c*) 2000 (*d*) 2001

Q. 22 NAVSTAR belongs to which country:

(*a*) India (*b*) Russia

(*c*) USA (*d*) China

Q. 23 GLONAS was launched in the year:

(*a*) 1982 (*b*) 1960

(*c*) 1992 (*d*) 2000

Q. 24 GLONAS belongs to which country:

(*a*) India (*b*) China

(*c*) Russia (*d*) USA

Q. 25 Gallileo will be launched in the year:

(*a*) 2008 (*b*) 2009

(*c*) 2010 (*d*) 2012

Q. 26 Gallileo will be owned by which country:

(*a*) European Union (*b*) Russia

(*c*) India (*d*) All

Q. 27 Data from Gallileo satellite will be shared by which of the following GPS satellite:

(*a*) NAVSTAR (*b*) GLONAS

(*c*) GAGAN (*d*) None

Q. 28 Which of the following GPS satellite is planned to be launched by India:

(*a*) NAVSTAR (*b*) GLONAS

(*c*) GAGAN (*d*) None

Q. 29 How many monitoring stations are there in NAVSTAR system:

(*a*) 5 (*b*) 6

(*c*) 9 (*d*) 10

Q. 30 How many ground antennas are there in NAVSTAR GPS SYSTEM:

(*a*) 3 (*b*) 5

(*c*) 8 (*d*) 10

Q. 31 How many master control stations are there for NAVSTAR system:

(*a*) 1 (*b*) 3

(*c*) 5 (*d*) 8

Q. 32 GPRS stands for:

(*a*) Global Pocket radio system

(*b*) Global pocket radar system

(*c*) Global positioning system

(*d*) None

Q. 33 GPRS is used in:

(*a*) Radio (*b*) Television

(*c*) Mobile (*d*) None

Q. 34 Each GPS satellite is provided with how many accurate clocks:

(*a*) 1 (*b*) 2

(*c*) 3 (*d*) 4

Q. 35 GPS instrument making company is:

(*a*) Trimble (*b*) Garmin

(*c*) Leica (*d*) All

Q. 36 At which of the following signal navigation messages are received:

(*a*) 10 Htz (*b*) 12 Htz

(*c*) 15Htz (*d*) 50 Htz

Q. 37 GPS receiver consists which of ther following carrier frequencies:

(*a*) L1 (*b*) L2

(*c*) L3 (*d*) L1 & L2

Q. 38 L1 frequency of GPS receiver works on:

(*a*) 1575.4 MHtz (*b*) 1227.6 MHtz

(*c*) Both (*d*) None

Q. 39 L2 frequency of GPS receiver works on:

(*a*) 1575.4 MHtz (*b*) 1227.6 MHtz

(*c*) Both (*d*) None

Q. 40 L1 frequency modulated with which of the following codes:

(*a*) P (*b*) C/A

(*c*) Both (*a*) and (*b*) (*d*) None

Q. 41 Which of the following person can use the P code signal:

(*a*) Millitary (*b*) Civillian

(*c*) Watershed mapper (*d*) Jhum mapper

Q. 42 Which of the following person can use the P code signal:

(*a*) Millitary (*b*) Civillian

(*c*) Watershed mapper (*d*) Jhum mapper

Q. 43 P codes and C/A codes are:

(*a*) Pseudo random binary codes

(*b*) Random binary codes

(*c*) Pseudo random logarithmic code

(*d*) Random logarithmic code

Q. 44 How GPS system works:

(*a*) Triangulation method (*b*) Quadrilateral method

(*c*) Perpendicular method (*d*) Horizontal method

Q. 45 Outline Principle of GPS works on:

(*a*) Range (*b*) Position

(*c*) Timing (*d*) All

Q. 46 The components of selective availability is/are:

(*a*) Dither (*b*) Epsilon

(*c*) Both (*d*) None

Q. 47 DEM refers to:

(*a*) Digital elevation model

(*b*) Digital eclipe model

(*c*) Digital equatorial model

(*d*) Digital elevator model

Q. 48 Error imposed in broadcasting message is known as:

(*a*) Dither (*b*) Epsilon

(*c*) Both (*d*) None

Q. 49 The error imposed in cloack frequency in GPS is known as:

(*a*) Dither (*b*) Epsilon

(*c*) Both (*d*) None

Q. 50 Which type of GPS is more accurate

(*a*) Urban Planning (*b*) Survey

(*c*) Both (*d*) None

ANSWERS

1	(*a*)	14	(*d*)	27	(*a*)	40	(*c*)
2	(*d*)	15	(*b*)	28	(*c*)	41	(*a*)
3	(*d*)	16	(*a*)	29	(*a*)	42	(*b*)
4	(*d*)	17	(*c*)	30	(*a*)	43	(*a*)
5	(*d*)	18	(*a*)	31	(*a*)	44	(*a*)
6	(*b*)	19	(*c*)	32	(*a*)	45	(*d*)
7	(*a*)	20	(*a*)	33	(*c*)	46	(*c*)
8	(*c*)	21	(*a*)	34	(*d*)	47	(*a*)
9	(*c*)	22	(*c*)	35	(*d*)	48	(*b*)
10	(*a*)	23	(*a*)	36	(*d*)	49	(*a*)
11	(*a*)	24	(*c*)	37	(*d*)	50	(*b*)
12	(*b*)	25	(*a*)	38	(*a*)		
13	(*d*)	26	(*a*)	39	(*b*)		

GIS

Q. 1 The 3 "W's "of Geography is:

(*a*) What is where (*b*) Why is it there

(*c*) Why do I care (*d*) All

Q. 2 In GIS, Geography tells about:

(*a*) Location (*b*) Data

(*c*) Integrated components (*d*) All

Q. 3 In GIS, Information tells about:

(*a*) Location (*b*) Data

(*c*) Integrated components (*d*) All

Q. 4 In GIS, System tells about:

(*a*) Location (*b*) Data

(*c*) Integrated components (*d*) All

Q. 5 A GIS contains:

(*a*) Input (*b*) Database

(*c*) Analysis (*d*) Output

(*e*) All

Q. 6 Which of the following is components of a GIS system:

(*a*) Maps (*b*) Information

(*c*) People (*d*) all

Q. 7 Which of the following is components of a GIS system:

(*a*) Software (*b*) Hardware

(*c*) Procedure (*d*) All

Q. 8 Spatially distributed features, activities or event, which are definable in space can be categorised as:

(*a*) Points (*b*) Lines

(*c*) Areas (*d*) All

Q. 9 Which of the following represents features that have a known location on earth:

(*a*) Spatial Data (*b*) Attribute Data

(*c*) Data Layers (*d*) Layer Type

(*e*) Topology

Q. 10 The information linked to the geographic features (spatial data) that describe those features is:

(*a*) Spatial Data (*b*) Attribute Data

(*c*) Data Layers (*d*) Layer Type

(*e*) Topology

Q. 11 The result of combining spatial and attribute data. Essentially adding the attribute database to the spatial location is done by:

(*a*) Spatial Data (*b*) Attribute Data

(*c*) Data Layers (*d*) Layer Type

(*e*) Topology

Q. 12 A layer type refers to the way spatial and attribute information are connected. There are two major layer types, vector and raster and they are called as:

(*a*) Spatial Data (*b*) Attribute Data

(*c*) Data Layers (*d*) Layer Type

(*e*) Topology

Q. 13 How geographic features are related to one another and where they are in relation to one another is known by:

(*a*) Spatial Data (*b*) Attribute Data

(*c*) Data Layers (*d*) Layer Type

(*e*) Topology

Q. 14: Spatial or coordinate data represents features that have a known location on the earth in X & Y locations:

(*a*) Points (*b*) Polygon

(*c*) Both (*d*) None

Q. 15. Spatial or coordinate data represents features that have a known location on the earth Connected X & Y locations that contain attribute information:

(*a*) Points (*b*) Polygon

(*c*) Both (*d*) None

Q. 16 Row and column matrix represent geographic space:

(*a*) Raster (*b*) Points

(*c*) Polygon (*d*) All

Q. 17 What Refers to the creation of digital spatial data:

(*a*) Data Input (*b*) Data Management

(*c*) Data Analysis (*d*) Data Output

Q. 18 What Refers to unique issues in the maintenance of spatial data such as error or level of accuracy; storing data; retrieving data; and metadata:

(*a*) Data Input (*b*) Data Management

(*c*) Data Analysis (*d*) Data Output

Q. 19 What allows users to answer questions that may not be explicitly stated in the data:

(*a*) Data Input (*b*) Data Management

(*c*) Data Analysis (*d*) Data Output

Q. 20 Refers to the method used to visually display analysis performed using GIS. Output can be in the form of *.*jpg* to large plotted images:

(*a*) Data Input (*b*) Data Management

(*c*) Data Analysis (*d*) Data Output

Q. 21 A GIS is closely related to which of the following sciences:

(*a*) Cartography (*b*) Remote sensing

(*c*) Computer aided designs (*d*) All

Q. 22 Non geographic spatial data information consists of:

(*a*) CAC (*b*) CAD

(*c*) CAM (*d*) All

Q. 23 The data which are on graphic information associated with a point line or area or elements in a GIS is known as:

(*a*) Attribute data (*b*) Raster data

(*c*) Both (*d*) None

Q. 24 A general photograph before scanning for computer analysis in general is in the form of:

(*a*) Attribute (*b*) Vector

(*c*) Rater (*d*) None

Q. 25 How a GIS works:

(*a*) Location (*b*) Condition

(*c*) Trend (*d*) All

Q. 26 What is the basic working philosophy of a GIS:

(*a*) Relating information from different sources

(*b*) Data capture

(*c*) Data integration

(*d*) All

Q. 27 What is the basic working philosophy of a GIS:

(*a*) Projection and registration

(*b*) Data structure

(*c*) Data integration and modeling

(*d*) All

Q. 28 Which of the following information can be included in attribute table:

(*a*) Soil pH (*b*) Soil type

(*c*) Soil texture (*d*) All

Q. 29 Data input in GIS can be made in the form of:

(*a*) Vector data (*b*) Raster data

(*c*) Alpha numeric data (*d*) All

Q. 30 Scanned map/images data in general can be put in a GIS system in the form of:

(*a*) Vector data (*b*) Raster data

(*c*) Alpha numeric data (*d*) All

Q. 31 Which of the following image/map is more accurate in calculation of an area of a feature:

(*a*) Vector data (*b*) Raster data

(*c*) Alpha numeric data (*d*) All

ANSWERS

1	(*d*)	9	(*a*)	17	(*a*)	25	(*d*)
2	(*a*)	10	(*b*)	18	(*b*)	26	(*d*)
3	(*b*)	11	(*c*)	19	(*c*)	27	(*d*)
4	(*c*)	12	(*d*)	20	(*d*)	28	(*d*)
5	(*e*)	13	(*e*)	21	(*d*)	29	(*d*)
6	(*d*)	14	(*a*)	22	(*d*)	30	(*b*)
7	(*d*)	15	(*b*)	23	(*a*)	31	(*d*)
8	(*d*)	16	(*a*)	24	(*b*)		

Chapter 7
Soil Informatics

Q. 1 The precision farming data base generally includes which of the following?

(*a*) Crop characteristics like, stage of the crop, crop health, nutrient requirement, etc.

(*b*) Detailed soil layer with physical and chemical properties, depth, texture, nutrient status, salinity

(*c*) Toxicity, soil temperature, productivity potential, etc.

(*d*) All the above

Q. 2 "Precision Agriculture" is an advanced e-agriculture application. It makes use of which of the following major components of technology?

(*a*) Geographical Information Systems (GIS) for analysis and management of spatial data and mapping

(*b*) Remote Sensing (RS) to identify

(*c*) Global Positioning Systems (GPS) to locate and define spatial features or activities that contributes to the quality of site-specific practices

(*d*) Variable Rate Technology (VRT) allowing targeted, site-specific input applications

(*e*) Yield monitoring for recording crop productivity as an historical database for crop management

(*f*) All the above

Q. 3 Which of the following is/are the application of computer technology?

(*a*) Prioritization of macro/micro watersheds for implementation and impact assessment of watershed through GIS and Remote Sensing (projects at national, state, district, taluka and levels. Projects already being carried out under DST)

(*b*) Forecasting of outbreak of pests and diseases based on soil water status and plant stress indicators in important crops such as rice, wheat, sugarcane, cotton, chilli, , pigeon pea, tea etc.

(*c*) Development of decision support system for precise management of resources at farm level at least in commercial/fruit/flower crops to begin with

(*d*) All the above

Q. 4 Which of the following is/are the application of computer technology?

(*a*) Soil mapping at cadastral scale using high resolution spatial, spectral and radiometric resolutions

(*b*) Quantification of soil loss

(*c*) Detection of water logging due to rising ground water table

(*d*) All the above

Q. 5 Which of the following is/are the application of computer technology?

(*a*) Delineation of salt-affected soils in black soil and sandy regions

(*b*) Soil moisture estimation and mapping using micro wave/optical/thermal remote sensing techniques in surface and root zone depth

(*c*) Land surface temperature estimation using thermal and microwave remote sensing techniques

(*d*) All the above

Q. 5 Which of the following is/are the application of computer technology?

(*a*) Hyper spectral studies on soils to establish quantitative relationship between spectral reflectance and soil properties

(*b*) Development of digital techniques for a variety of applications using GIS techniques for e.g., soil suitability to crops, land capability classification and land irrigability assessment etc.

(*c*) Yield models by integration of biophysical simulation and regional level crop models

(*d*) All the above

Q. 6 Which of the following is/are the application of computer technology?

(*a*) Land information system

(*b*) Market intelligent

(*c*) Pest management system

(*d*) All the above

Q. 7 Which of the following can be a suitable area for computer application?

(*a*) Identification of crops and estimation of area and production of short duration crops grown in fragmented land holdings

(*b*) Forecasting of droughts/floods

(*c*) Detection of crop stress due to nutrients, pests and diseases and quantification of their effects on crop yield

(*d*) Automation of land evaluation procedures for a variety of applications using GIS techniques

(*e*) All the above

Q. 8 Which of the following can be a suitable area for computer application?

(*a*) Information on sub-surface soil horizons

(*b*) Extending precision farming database to smaller farm size and/or diverse crops/cropping systems (*c*) Developing decision support systems for management of biotic and abiotic stresses at the farm level

(*d*) More accurate yield models

(*e*) All the above

Q. 9 Which of the following can be a suitable area for computer application?

(*a*) Estimation of depth of water in reservoirs and quality assessment of ground water

(*b*) Better than 1m contours for watershed development plan at the micro level

(*c*) Use of remote sensing and precision farming technologies in intercropping/multiple cropping situations

(*d*) Identifying ways and means of reducing the cost of RS, GIS and precision farming technologies and time gap in collection, interpretation and dissemination of data to enable their usage on a large scale

(*e*) All the above

Q. 10 Which of the following can be a suitable area for computer application?

(*a*) Convincing evidence to prove the utility and economic viability of these technologies so as to mobilize support for R and D work

(*b*) Human resource development to hasten the process of large scale use of unexplained and cutting edge technologies that have tremendous scope and potential

(*c*) Both (*a*) and (*b*)

(*d*) None of the above

Q. 11 Information technology which is one of the most promising technologies of the present era, can be utilized for this very purpose using which of the following?

(*a*) Information support systems

(*b*) Decision support systems

(*c*) Expert support systems

(*d*) All the above

Q. 12 Which of the following can be a suitable area for computer application?

(*a*) The crop forecasting, input management command

(*b*) Watershed management

(*c*) Land and water resources development

(*d*) All the above

Q. 13 Which of the following can be a suitable area for computer application?

(*a*) Drinking water potential mapping

(*b*) Precision management natural disaster management

(*c*) Fishery management

(*d*) Hill area development and post harvest management

(*e*) All the above

Q. 14 The growers/land users urgently information about the:

(*a*) Soil attribute space: Soil fertility data

(*b*) Soil functions: Soil affects crop production factor, also affecting thereby many alternative land use systems

(*c*) Soil land use distribution: The georeferenced occurrence of the soils/lands can be inventorized and disseminated to various land users

(*d*) Soil-factor relationships: Soil genetics can forecast future processes of soils induced by local and global changes of systems

(*e*) All the above

Q. 15 The SIS has to support following which of the following principles?

(*a*) Identification of data and methods

(*b*) Data selection, querying and data evaluation

(*c*) Disposition of results including display and documentation

(*d*) All the above

Q. 16 Soil Information Systems are needed to:

(*a*) Establish the potentials and limitations of different kinds of soils for crop production

(*b*) Keep track of different forms of soil degradation problems like acidification, jhuming, erosions, pollutions etc.

(*c*) To establish the circumstances which lead different land use systems, soil and land use productivity can be maintained or improved and the situations leading to a particular kind of practices these are degraded

(*d*) To serve as a basis of transfer of technology and

(*e*) To aid in decision making level

(*f*) All the above

Q. 17 To provide the necessary data for improved mapping, modeling and monitoring of changes of world soil and terrain resources. Which of the following SIS is used?

(*a*) SOTER (*b*) HUMSOTER

(*c*) LASOTER (*d*) RAMSOTER

Q. 18 'HunSOTER' describes which country's soil information system?

(*a*) Hungry (*b*) Poland

(*c*) USA (*d*) England

Q. 19 'RAMSOTER-200' describes which country's soil information system?

(*a*) Hungry (*b*) Poland

(*c*) USA (*d*) Romania

Q. 20 'LASOTER' describes which country's soil information system?

(*a*) Hungry (*b*) Latin America

(*c*) USA (*d*) Romania

Q. 21 'FISBOBGR' describes which country's soil information system?

(*a*) Germany (*b*) Latin America

(*c*) USA (*d*) Romania

Q. 22 'NIBIS' describes which country's soil information system?

(*a*) Germany (*b*) Latin America

(*c*) USA (*d*) Romania

Q. 23 'HESBIS' describes which country's soil information system?

(*a*) Germany (*b*) Latin America

(*c*) USA (*d*) Romania

Q. 24 'SOPIC' describes which country's soil information system?

(*a*) Germany (*b*) Latin America

(*c*) USA (*d*) Romania

Q. 25 'PUGIS' describes which country's soil information system?

(*a*) Germany (*b*) Czech Republic

(*c*) USA (*d*) Romania

Q. 26 'NIJOS' describes which country's soil information system?

(*a*) Germany (*b*) Czech Republic

(*c*) Norway (*d*) Romania

Q. 27 'ESPLAN' describes which country's soil information system?

(*a*) Germany (*b*) Czech Republic

(*c*) Norway (*d*) Italy

Q. 28 'Land IS' describes which country's soil information system?

(*a*) Germany (*b*) UK

(*c*) Norway (*d*) Italy

Q. 29 'LIS' describes which country's soil information system?

(*a*) Germany (*b*) UK

(*c*) Norway (*d*) Greece

Q. 30 'ISRIC' describes which country's soil information system?

(*a*) The Netherlands (*b*) UK

(*c*) Norway (*d*) Greece

Q. 31 'ILWIS' describes which country's soil information system?

(*a*) The Netherlands (*b*) UK

(*c*) Norway (*d*) Greece

Q. 32 'RSLIS' describes which country's soil information system?

(*a*) The Netherlands (*b*) Romania

(*c*) Norway (*d*) Greece

Q. 33 'NASIS' describes which country's soil information system?

(*a*) The Netherlands (*b*) Romania

(*c*) USA (*d*) Greece

Q. 34 'CRIES' describes which country's soil information system?

(*a*) The Netherlands (*b*) Romania

(*c*) USA (*d*) Greece

Q. 35 'ARS' describes which country's soil information system?

(*a*) Australia (*b*) Romania

(*c*) USA (*d*) Greece

Q. 36 'UMSIS' describes which country's soil information system?

(*a*) Australia (*b*) Romania

(*c*) China (*d*) Greece

Q. 37 'PLANTAGRO' describes which country's soil information system?

(*a*) Australia (*b*) Thailand

(*c*) China (*d*) Greece

Q. 38 'NRIS' describes which country's soil information system?

(*a*) India (*b*) Thailand

(*c*) China (*d*) Greece

Q. 39 Some of the other applications of SIS can be:

(*a*) Assessment of soil and land productivity potential and constraints

(*b*) Assessment of soil quality

(*c*) Soil and land use planning

(*d*) Use of automated land evaluation System

(*e*) All the above

Q. 40 Some of the other applications of SIS can be:

(*a*) Soil and Land irrigability and water management

(*b*) Use of Soil Database in Watershed Management

(*c*) Estimation of pedotransfer functions from SIS

(*d*) Physical rating index, Soil Physical Constraints and Soil Management Recommendations

(*e*) All the above

Q. 41 Some of the other applications of SIS can be:

(*a*) Tillage rating of soils and developing Tillage Guide

(*b*) Root zone Capacity mapping

(*c*) Integrated nutrient management

(*d*) Neural network approach to monitor heavy metal contamination

(*e*) All the above

Q. 42 Future strategies for development of Soil and Land information systems and its application is/are

(*a*) Build up of national Soil and Terrain Resources Database

(*b*) Building up of Soil database at State/District/Block level for making a national network

(*c*) Requirement of database processing, evaluation and interpretation

(*d*) Establishment of Long term Monitoring Plots

(*e*) All the above

Q. 43 Future strategies for Development of Soil and Land information systems and its application is/are:

(*a*) Launching of a website

(*b*) Use of Fuzzy rules and GIS in soil prediction

(*c*) Linking of digital soil maps and databases to simulation models

(*d*) Challenges of local soil management

(*e*) All the above

Q. 44 Soil Information Systems are needed to

(*a*) Establish the potentials and limitations of different kinds of soils for crop production

(*b*) Keep track of different forms of soil degradation problems like acidification, jhumings, erosions, pollutions etc.

(*c*) To establish the circumstances which lead different land use systems, soil and land use productivity can be maintained or improved and the situations leading to a particular kind of practices these are degraded

(*d*) All the above

Q. 45 Soil Information Systems are needed to:

(*a*) To serve as a basis of transfer of technology

(*b*) To aid in decision making level

(*c*) Both (*a*) and (*b*)

(*d*) None of the above

Q. 46 CRES (Crop Environment Resource Synthesis) has developed DSSIFER model for:

(*a*) Fertilizer recommendation (*b*) Maize

(*c*) Groundnut (*d*) Wheat

Q. 47 Geographical information systems (GIS) is a process to:

(*a*) Input (*b*) Store

(*c*) Retrieve (*d*) Analyze

(*e*) Display geographic data (*f*) All the above

Q. 48 Why robotics is used in agriculture?

(*a*) Require intelligence and quick decisions

(*b*) Increasing Market for high quality products

(*c*) Both (*a*) and (*b*)

(*d*) All the above

Q. 49 Biological objects' characteristics in relation to robotics:

(*a*) Physical – shape, size, mass, texture

(*b*) Dynamic – Friction (for gripping the plants, animals etc.)

(*c*) Optical – reflectance (weeds, flowers with honey, male insects' identification etc.)

(*d*) Sonic- vibration (differentiation of watery and tissue objects)

(*e*) Breathing activity of harvested objects

(*f*) All the above

Q. 50 Agricultural imaging is:

(*a*) Selective harvesting

(*b*) Machine guidance

(*c*) Plant identification

(*d*) Plant health assessment

(*e*) All the above

Q. 51 The precision farming data base generally includes which of the following?

(*a*) Microclimate data (seasonal and daily) about the canopy temperature, wind direction and speed, humidity etc.

(*b*) Surface and sub surface drainage conditions

(*c*) Irrigation facilities, water availability, and other planning inputs of interest

(*d*) All the above

Q. 52 Decision support systems are basically:

(*a*) Interactive computerized tools under user control that help in making decisions (*b*) Have analytical capabilities (graphical and modeling)

(*c*) Allow combining computer input and output with judgment to produce meaningful information

(*d*) Are custom fit

(*e*) All the above

Q. 53 Objectives of decision support systems is/are:

(*a*) Insights to key issues and problems in decision type information systems

(*b*) Understand the level in decision hierarchy

(*c*) Know the different solution approaches

(*d*) All the above

Q. 54 DSS are required in agriculture because of which of the following?

(*a*) Right decisions are important from both economic and environmental points of view

(*b*) There is uncertainty in weather and markets

(*c*) Decision making is complex

(*d*) Available information is incomplete

(*e*) All the above

Q. 55 Who needs a DSS?

(*a*) Farmers (*b*) Extension advisers

(*c*) Development departments (*d*) Scientists

(*e*) All the above

Q. 56 Who needs a DSS?

(*a*) Planners (*b*) Consultants

(*c*) Insurance companies (*d*) Industry

(*e*) All the above

Q. 57 What type of decisions is required for making DSS?

(*a*) Selection of crops and cropping systems

(*b*) Crop management

(*c*) Variety (*d*) Irrigation

(*e*) All the above

Q. 58 What type of decisions is required for making DSS?

(*a*) Fertilizer

(*b*) Pest control

(*c*) Natural resource management

(*d*) All

Q. 59 What type of decisions is required for making DSS?

(*a*) Farm management

(*b*) Watershed/command area/regional management

(*c*) Land use planning

(*d*) All the above

Q. 60 What are the characteristics of these decisions?

(*a*) Objective/Structured

(*b*) Subjective/Unstructured

(*c*) Both (*a*) and (*b*)

(*d*) None of the above

Q. 61 BASIC acronym stands for:

(*a*) Beginner's All-purpose Symbolic Instruction Code

(*b*) Begin all simple coding

(*c*) Begin assemble simulate instruct code

(*d*) None of the above

Q. 62 BASIC computer program uses which of the following?

(*a*) Words (*b*) Symbols

(*c*) Numbers (*d*) All the above

Q. 63 Kinds of computer languages, include which of the following?

(*a*) Pascal (*b*) C++

(*c*) FORTRAN (*d*) COBOL

(*e*) BASIC (*f*) All the above

Q. 64 Older versions of DOS (pre-5.0) came with a version of BASIC is called as:

(*a*) GW-BASIC (*b*) MW-BASIC

(*c*) CW-BASIC (*d*) GM-BASIC

Q. 65 Newer versions of DOS come with:

(*a*) QBasic (*b*) G-BASIC

(*c*) M-BASIC (*d*) C-BASIC

Q. 66 User interface objects in a computer programme include which of the following?

(*a*) Buttons (*b*) Menus

(*c*) Text boxes (*d*) All the above

Q. 67 ICT means:

(*a*) Information and communication technology

(*b*) Information for computer transfer

(*c*) Intelligent computing technology

(*d*) All the above

Q. 68 Computer network examples includes:

(*a*) LAN (*b*) WAN

(*c*) MAN (*d*) All the above

Q. 69 A modem is required when connectivity is through:

(*a*) Telephone (*b*) Cable

(*c*) Leased line (*d*) All the above

Q. 70 Netscape Navigator, Internet Explorer and Lynx are:

(*a*) Browser software (*b*) Internet account

(*c*) Modem (*d*) None of the above

Q. 71 Which of the following is a search engine?

(*a*) google (*b*) yahoo.com

(*c*) msn.com (*d*) All the above

Q. 72 Boolean logical operators are:

(*c*) AND (*b*) OR

(*c*) NOT (*d*) All the above

Q. 73 Ikisan is an initiative of which organization?

(*a*) ITC (*b*) Nagarjuna group

(*c*) Chambal group (*d*) None of the above

Q. 74 Village Knowledge centre is an initiative of which organization?

(*a*) MSSRF (*b*) Tata social welfare

(*c*) ITC (*d*) None of the above

Q. 75 Who invented the first calculating machine?

(*a*) Abacus (*b*) Bebetus

(*c*) Babbage (*d*) None of the above

Q. 76 The permanent memory of a computer is stored in:

(*a*) ROM (*b*) RAM

(*c*) Both (*a*) and (*b*) (*d*) None of the above

Q. 77 The first system software to run on a computer was:

(*a*) DOS (*b*) Macintosh

(*c*) Windows (*d*) None of the above

Q. 78 Monitor of a computer can be a:

(*a*) Monologue (*b*) Digital

(*c*) Both (*a*) and (*b*) (*d*) None of the above

Q. 79 Basic component of a computer network include:

(*a*) Modems (*b*) VSAT

(*c*) Routers (*d*) Hubs

(*e*) All the above

Q. 80 Page maker is a:

(*a*) Text processing software

(*b*) Graphics software

(*c*) Both (*a*) and (*b*)

(*c*) None the above

Q. 81. A database management (DBMS) package software is

(*a*) FOXPRO (*b*) ORACLE

(*c*) REFLEX (*d*) MS-ACCESS

(*e*) All the above

Q. 82 Which of the following spreadsheets can also be used as a data storage, retrieval and manipulation package?

(*a*) Lotus 1-2-3 (*b*) QUATROPRO

(*c*) MULTIPLAN (*d*) MS EXCELL

(*e*) All the above

Q. 83 Which of the following is a statistical software package?

(*a*) MSTATC (*b*) MATLAB

(*c*) MINITAB (*d*) SAS

(*e*) All the above

Q. 84 Transmission media for a network may be of which type?

(*a*) Baseband signalling

(*b*) Broadband signalling

(*c*) Both (*a*) and (*b*)

(*d*) None of the above

Q. 85 Which of the following is a network type example?

(*a*) Peer to peer (*b*) Hierarchical

(*c*) Both (*a*) and (*b*) (*d*) None of the above

Q. 86 Every internet email has at least how many parts?

(*a*) One (*b*) Two

(*c*) Three (*d*) None of the above

Q. 87 The domain name in an internet email is:

(*a*) Mail server address

(*b*) Different organizations

(*c*) Both (*a*) and (*b*)

(*d*) None of the above

Q. 88 Archie is a search program that will look through different:

(*a*) Files (*b*) Data

(*c*) Both (*a*) and (*b*) (*d*) None of the above

Q. 89 Veronica is a:

(*a*) Gopher tool (*b*) Ftp tool

(*c*) Both (*a*) and (*b*) (*d*) None of the above

Q. 90 Basic component of database activities includes:

(*a*) Database management system

(*b*) Database administrator

(*c*) Both (*a*) and (*b*)

(*d*) None of the above

Q. 91 Which of the following was the first computer?

(*a*) ENIAC (*b*) Univac

(*c*) Both (*a*) and (*b*) (*d*) None of the above

Q. 92 The smallest data element in a database management system is:

(*a*) Field (*b*) Data element

(*c*) File (*d*) None of the above

Q. 93 Functions of the a database administrator includes:

(*a*) Creation of the database

(*b*) Performance optimization

(*c*) Data protection, authorization

(*d*) All the above

Q. 94 Which of the following is a major database CDROM in agriculture?

(*a*) CABCD (*b*) ASFA

(*c*) VETCD (*d*) RURAL

(*e*) All the above

Q. 95 SPAR1 is a statistical package developed at:

(*a*) IARI (*b*) IASRI

(*c*) ISI (*d*) None of the above

Q. 96 The first generation computers used which of the following technology?

(*a*) Thermionic valve

(*b*) Transistor technology

(*c*) Microprocessor based technology

(*d*) Chip technology

Q. 97 Contemporary computing is based on which of the following technology?

(*a*) Thermionic valve

(*b*) Transistor technology

(*c*) Microprocessor based technology

(*d*) Chip technology

Q. 98 Which of the following is a type of computer virus?

(*a*) Boot sector (*b*) File infector

(*c*) Trojan horse (*d*) File destroyer

(*e*) All the above

Q. 99 Which of the following is a GIS package?

(*a*) ARCINFO (*b*) PAMAP

(*c*) MAPINFO (*d*) GRASS

(*e*) All the above

Q. 100 Which of the following is a programme for creating the mail message in email?

(*a*) Eudora (*b*) Pine

(*c*) Window (*c*) Telix

(*e*) All the above

Q. 101 Which of the following is an effective email transfer agent?

(*a*) SMTP (*b*) POP

(*c*) IMAP (*d*) UUCP

(*e*) All the above

Q. 102 DSSAT (Decision Support System for Agro'echnology Transfer) was developed by:

(*a*) IBSNAT (*b*) NRSA

(*c*) ARS (*d*) None of the above

Q. 103 CRES (Crop Environment Resource Synthesis) has developed CERES model for:

(*a*) Rice (*b*) Maize

(*c*) Sorghum (*d*) All the above

Q. 104 CRES (Crop Environment Resource Synthesis) has developed PNUTGRO model for:

(*a*) Rice (*b*) Maize

(*c*) Groundnut (*d*) Wheat

Q. 105 CRES (Crop Environment Resource Synthesis) has developed SOYGRO model for:

(*a*) Rice (*b*) Soybean

(*c*) Groundnut (*d*) Wheat

ANSWERS

1	(*d*)	23	(*a*)	45	(*c*)	67	(*d*)
2	(*f*)	24	(*a*)	46	(*c*)	68	(*d*)
3	(*d*)	25	(*b*)	47	(*f*)	69	(*d*)
4	(*d*)	26	(*c*)	48	(*c*)	70	(*a*)
5	(*d*)	27	(*d*)	49	(*e*)	71	(*d*)
6	(*d*)	28	(*b*)	50	(*e*)	72	(*d*)
7	(*d*)	29	(*b*)	51	(*e*)	73	(*b*)
8	(*e*)	30	(*a*)	52	(*e*)	74	(*a*)
9	(*e*)	31	(*a*)	53	(*d*)	75	(*a*)
10	(*e*)	32	(*a*)	54	(*e*)	76	(*a*)
11	(*c*)	33	(*c*)	55	(*e*)	77	(*a*)
12	(*d*)	34	(*a*)	56	(*e*)	78	(*c*)
13	(*e*)	35	(*a*)	57	(*d*)	79	(*e*)
14	(*e*)	36	(*c*)	58	(*d*)	80	(*a*)
15	(*d*)	37	(*b*)	59	(*d*)	81	(*e*)
16	(*f*)	38	(*a*)	60	(*c*)	82	(*e*)
17	(*a*)	39	(*e*)	61	(*a*)	83	(*e*)
18	(*a*)	40	(*e*)	62	(*d*)	84	(*c*)
19	(*d*)	41	(*e*)	63	(*f*)	85	(*c*)
20	(*b*)	42	(*e*)	64	(*a*)	86	(*c*)
21	(*a*)	43	(*e*)	65	(*a*)	87	(*a*)
22	(*a*)	44	(*d*)	66	(*d*)	88	(*a*)

89	(*a*)	94	(*e*)	98	(*e*)	102	(*a*)
90	(*c*)	95	(*b*)	99	(*e*)	103	(*d*)
91	(*a*)	96	(*a*)	100	(*e*)	104	(*c*)
92	(*a*)	97	(*d*)	101	(*e*)	105	(*b*)
93	(*d*)						

Chapter 8

Basics of Soil Science

Q. 1 Science that deals with soil genesis and classification is called as:

(*a*) Petrology (*b*) Pedology

(*c*) Pedagogy (*d*) Petrography

Q. 2 Rocks formed by cooling of the molten materials (magma) from the sun is called as:

(*a*) Metamorphic rocks (*b*) Igneous rocks

(*c*) Sedimentary rocks (*d*) All

Q. 3 Of the followings, which is/are the examples of igneous rocks?

(*a*) Granites (*b*) Basalts

(*c*) Syenites (*d*) All

Q. 4 Rocks formed from igneous rocks by the influence of high pressure and temperature are called as:

(*a*) Metamorphic rocks (*b*) Primary rocks

(*c*) Sedimentary rocks (*d*) All

Q. 5 Gneiss, Schist and Slate are the examples of:

(*a*) Metamorphic rocks (*b*) Sedimentary rocks

(*c*) Primary rocks (*d*) Igneous rocks

Q. 6 Sandstone and limestone are:

(*a*) Metamorphic rocks (*b*) Sedimentary rocks

(*c*) Primary rocks (*d*) Igneous rocks

Q. 7 Major component of igneous rocks are:

(*a*) Quartzs

(*b*) Amphiboles and pyroxenes

(*c*) Feldspars

(*d*) Micas

Q. 8 Percentage of quartz in igneous rocks is:

(*a*) 4 (*b*) 12

(*c*) 18 (*d*) 60

Q. 9 Percentage of feldspar in igneous rocks is:

(*a*) 4 (*b*) 12

(*c*) 18 (*d*) 60

Q. 10 Percentage of mica in igneous rocks is:

(*a*) 4 (*b*) 12

(*c*) 18 (*d*) 60

Q. 11 The process of disintegration and decomposition of minerals within the soil is called as:

(*a*) Geochemical weathering

(*b*) Pedochemical weathering

(*c*) Physical weathering

(*d*) Chemical weathering

Q. 12 The process of weathering that takes place before soil development and continues beneath the soil cover is called as:

(*a*) Geochemical weathering

(*b*) Pedochemical weathering

(*c*) Physical weathering

(*d*) Chemical weathering

Q. 13 Of the following, which is an important chemical weathering process?

(*a*) Carbonation (*b*) Oxidation

(*c*) Hydrolysis (*d*) Hydration

Q. 14 Which is/are the chemical weathering processes?

(*a*) Carbonation (*b*) Hydrolysis

(*d*) Hydration (*d*) All

Q. 15 Which of the following processes is responsible for the formation of serpentine from olivine?

(*a*) Hydrolysis (*b*) Hydration

(*c*) Oxidation (*d*) Reduction

Q. 16 Which of the following processes is responsible for the formation of gypsum from calcium sulphate?

(*a*) Hydration (*b*) Carbonation

(*c*) Oxidation (*d*) Hydrolysis

Q. 17 Which of the following processes is responsible for the formation of allophane from orthoclase feldspar?

(*a*) Hydration (*b*) Carbonation

(*c*) Oxidation (*d*) Hydrolysis

Q. 18 Transformation of biotite to kaolinite takes place by the process of:

(*a*) Physical weathering (*b*) Chemical weathering

(*c*) Biological weathering (*d*) Both b and c

Q. 19 Weathered products of rocks and rock minerals are called as:

(*a*) Pedon (*b*) Parent materials

(*c*) Solum (*d*) Polypedons

Q. 20 Most easily weatherable mineral is:

(*a*) Gypsum (*b*) Biotite

(*c*) Olivine (*d*) Calcite

Q. 21 Mineral formed by exfoliation, a type of physical weathering, is:

(*a*) Granite (*b*) Basalt

(*c*) Gypsum (*d*) Calcite

Q. 22 Stability of minerals to weathering and to decomposition is related to:

(*a*) Degree of basicity

(*b*) Degree of linkage to tetrahedron

(*c*) Silica: Sesquioxide ratio

(*d*) All

Q. 23 When bases are more, the stability of minerals will:

(*a*) Decrease (*b*) Increase

(*c*) Unaltered (*d*) None

Q. 24 The main products of weathering is/are:

(*a*) Silica (*b*) Alumina

(*c*) Soluble cations (*d*) All

Q. 25 Secondary minerals that act as a primary minerals are:

(*a*) Biotite and muscovite

(*b*) Muscovite and montmorillonite

(*c*) Mica and biotite

(*d*) Mica and muscovite

Q. 26 Which is the correct sequence of formation of the secondary clay minerals?

(*a*) Mica → Illite → Vermiculite → Montmorillonite

(*b*) Montmorillonite → Vermiculite → Illite → Mica

(*c*) Vermiculite → Montmorillonite → Illite → Mica

(*d*) Mica → Vermiculite → Montmorillonite → Illite

Q. 27 Presence of more calcium and magnesium are more conducive for the formation of:

(*a*) Chlorite (*b*) Montmorillonite

(*c*) Kaolinite (*d*) Illite

Q. 28 Which is the product of exhaustive hydrolysis and leaching?

(*a*) Vermiculite (*b*) Montmorillonite

(*c*) Illite (*d*) Kaolinite

Q. 29 Which is the dominant clay mineral in humid tropics?

(*a*) Kaolinite (*b*) Montmorillonite

(*c*) Vermiculite (*d*) Illite

Q. 30 Clay minerals predominate in the soils of arid regions are:

(*a*) Montmorillonite and vermiculite

(*b*) Illite and montmorillonite

(*c*) Illite and chlorite

(*d*) Illite and vermiculite

Q. 31 The natural aggregation of soil separates into peds is called as:

(*a*) Soil textural class (*b*) Soil separates

(*c*) Soil structure (*d*) Soil consistency

Q. 32 The structural type most common in the B horizon of soils of tropical regions is:

(*a*) Platy (*b*) Blocky

(*c*) Prism like (*d*) Granular

Q. 33 Soil permeability refers to:

(*a*) Stickiness and plasticity of soil

(*b*) Total pore space

(*c*) Movement of air and water through soil

(*d*) None

Q. 34 Soil consistency refers to:

(*a*) The degree of plasticity and stickiness of a soil

(*b*) Movement of air and water through soil

(*c*) The occurrence of similar soil profiles in similar landscape positions

(*d*) Total pore space

Q. 35 Incorporating crop residues into the soil:

(*a*) Improves soil structure (*b*) Improves soil texture

(*c*) Improves both of the above (*d*) None

Q. 36 Soil permeability is affected mainly by:

(*a*) Soil texture (*b*) Soil structure

(*c*) Soil texture and structure (*d*) Soil colour

Q. 37 Soil bulk density relates the oven dry weight of solids to total volume of soil. How is bulk density calculated?

(*a*) Weight of oven dry soil/Volume

(*b*) Volume/Weight of oven dry soil

(*c*) Weight of oven dry soil × Volume

(*d*) None of the above

Q. 38 If 100 cm^3 of soil has an oven dry weight of 135 grams, what is its bulk density?

(*a*) 13,500 g/cm^3 (*b*) 0.75 cm^3/g

(*c*) 1.35 g/cm^3 (*d*) 1.53 g/cm^3

Q. 39 Soil bulk density is greatest for:

(*a*) Sand (*b*) Clay

(*c*) Peat (*d*) Muck

Q. 40 Soil compaction results in (an) ______________ in soil bulk density:

(*a*) Increase (*b*) Decrease

(*c*) No change (*d*) None

Q. 41 A soil with a large total amount of pore space is always permeable (True/False):

(*a*) True (*b*) False

Q. 42 Soil colour is used as an indicator of:

(*a*) Soil texture (*b*) Structure

(*c*) Consistence (*d*) Drainage

Q. 43 Sandy loam is an example of:

(*a*) Soil texture (*b*) Soil structure

(*c*) Soil textural class (*d*) Soil

Q. 44 Which kind of soil parent material is not found in India?

(*a*) Transported (*b*) Residual

(*c*) Glacial outwash (*d*) None

Q. 45 The smallest volume that represents the range of characteristics of a soil in the landscape is:

(*a*) Ped (*b*) Pedon

(*c*) Order (*d*) Regolith

Q. 46 An example of a soil separate is:

(*a*) Sandy (*b*) Sand

(*c*) Loam (*d*) Silt loam

Q. 47 Soils named for locations, such as Hebbal and Nagapatinam, are examples of:

(*a*) Soil series (*b*) Soil orders

(*c*) Soil suborders (*d*) Great soil group

Q. 48 The soil order most common to the southeastern India is:

(*a*) Spodosols (*b*) Mollisols

(*c*) Ultisols (*d*) Aridisols

Q. 49 If a soil is classified as a typic fluvaquent, what is the soil order?

(*a*) Ultisols (*b*) Entisols

(*c*) Aridisols (*d*) Alfisols

Q. 50 The volume used to calculate soil bulk density includes

(*a*) The solid material and the pore space

(*b*) Density of the solid particles only

(*c*) Only the pore space

(*d*) Moisture percentage

Q. 51 Particle density is the

(*a*) Solid material and the pore space

(*b*) Density of the solid particles only

(*c*) Total pore space

(*d*) None of the above

Q. 52 If a 100 cm³ soil sample is saturated (all the pore space is filled with water) contains 40 grams of water, what is the percent pore space?

(*a*) 60% (*b*) 40%

(*c*) 30% (*d*) 50%

Q. 53 Gray colours throughout the B horizon of a soil profile are an indication of

(*a*) Well drained soil

(*b*) Very poorly drained soil

(*c*) Soil colour doesn't mean anything

(*d*) None

Q. 54 Which of the following is not a process of weathering which breaks down rocks?

(*a*) Wetting (*b*) Hammering

(*c*) Freezing (*d*) Thawing

Q. 55 $KAl_3(OH)_6(SO_4)_2$ is the structural formula of:

(*a*) Diorite (*b*) Syenite

(*c*) Alunite (*d*) Jarosite

Q. 56 Alunite mottles are found in:

(*a*) Melanic horizon (*b*) Salic horizon

(*c*) Sulphuric horizon (*d*) Anthropic horizon

Q. 57 What is the percentage of base saturation in sombric diagnostic endopedon?

(*a*) <30% (*b*) <40%

(*c*) <50% (*d*) <60

Q. 58 Presence of artifacts is a characteristic feature of:

(*a*) Anthropic epipedon (*b*) Plaggen epipedon

(*c*) Melanic epipedon (*d*) Mollic epipedon

Q. 59 Highest hierarchy in the classification of USDA soil classification is:

(*a*) Soil order (*b*) Soil suborder

(*c*) Great soil group (*d*) Soil family

Q. 60 Soil forming processes are used as primary criteria in the classification of:

(*a*) Soil order (*b*) Soil suborder
(*c*) Great soil group (*d*) Soil family

Q. 61 Granodiorite is a:

(*a*) Volcanic igneous rock (*b*) Metamorphic rock
(*c*) Plutonic igneous rock (*d*) Sedimentary rock

Q. 62 Black soils have high content of:

(*a*) R_2O_3 (*b*) Al_2O_3
(*c*) Fe_2O_3 (*d*) Silica

Q. 63 Primary form(s) of nitrogen in soil is/are:

(*a*) Organic (*b*) Inorganic
(*c*) Both (*a*) and (*b*) (*d*) All

Q. 64 Predominant form(s) of phosphorous in soil is/are:

(*a*) Organic (*b*) Inorganic
(*c*) Both (*a*) and (*b*) (*d*) All

Q. 65 Predominant form(s) of sulphur in soil is/are:

(*a*) Organic (*b*) Inorganic
(*c*) Both (*a*) and (*b*) (*d*) All

Q. 66 Primary inorganic form of phosphorous in soil is:

(*a*) Iron and aluminium phosphate minerals
(*b*) Apatite
(*c*) Augite
(*d*) Anorthite

Q. 67 What are the two main sources of inorganic sulphur in soil?

(*a*) Gypsum and Epsum
(*b*) Pyrites and Epsum
(*c*) Gypsum and Pyrites
(*d*) Gypsum and Schoenite

Q. 68 In the determination of carbon content, the carbon dioxide produced can be estimated by:

(*a*) Colorimetric method

(*b*) Walkley and Black method

(*c*) Wet combustion method

(*d*) Dry combustion method

Q. 69 Total carbon in soil is measured by:

(*a*) Colorimetric method

(*b*) Walkley and Black method

(*c*) Wet combustion method

(*d*) Dry combustion method

Q. 70 What is the percentage of organic carbon in organic matter?

(*a*) 48 (*b*) 58

(*c*) 78 (*d*) 68

Q. 71 Organic matter = Organic carbon × ______:

(*a*) 1.724 (*b*) 2.724

(*c*) 3.724 (*d*) 4.724

Q. 72 What is the C: N ratio of Indian soils?

(*a*) 30- 32: 1 (*b*) 100: 1

(*c*) 20- 22: 1 (*d*) 10 – 12: 1

Q. 73 Climatic conditions conducive for the accumulation of organic matter in soil are:

(*a*) Low temperature and low rainfall

(*b*) Low temperature and high rainfall

(*c*) High temperature and low rainfall

(*d*) High temperature and high rainfall

Q. 74 When the NSQ (Neiderslag Sattingungsdruck Quotient) is high, organic matter and nitrogen content in soil will be:

(*a*) High (*b*) Medium

(*c*) Low (*d*) Equal

Q. 75 The beneficial effect of organic matter is very high in:

(a) Fine textured soils

(b) Medium textured soils

(c) Coarse textured soils

(d) All

Q. 76 Humic acid is soluble in:

(a) Dilute acid (b) Dilute alkali

(c) Concentrated acid (d) Concentrated alkali

Q. 77 Precipitation of humic acid takes place in:

(a) Dilute acid (b) Dilute alkali

(c) Concentrated acid (d) Concentrated alkali

Q. 78 Compound(s) obtained from the precipitated humic acid when refluxed with ethanol is/are:

(a) Fulvic acid (b) Humin

(c) Hymatomelanic acid (d) Both (a) and (d)

Q. 79 Humin can be extracted by:

(a) 2N HCl (b) 2N NaOH

(c) 2N Na_2CO_3 (d) 4% H_2O_2

Q. 80 Insoluble fraction of organic matter is:

(a) Humin (b) Fulvic acid

(c) Humic acid (d) Hymatomelanic acid

Q. 81 Organic compound that contain more oxygen but less carbon and nitrogen is:

(a) Humic acid (b) Fulvic acid

(c) Humin (d) Hymatomelanic acid

Q. 82 Organic compound that contain less oxygen but more carbon and nitrogen is:

(a) Humic acid (b) Fulvic acid

(c) Humin (d) Hymatomelanic acid

Q. 83 Of the following, which is/are the more stable metal-organic interactions found in soil?

(*a*) Metal-humic acid (*b*) Metal-fulvic acid

(*c*) Metal-humin (*d*) Both (*a*) and (*b*)

(*e*) All the above

Q. 84 What is the upper limit of colloidal particles?

(*a*) 0.02 microns (*b*) 0.2 microns

(*c*) 2 microns (*d*) 0.1 microns

Q. 85 What is the cation exchange capacity of humus?

(*a*) 150–250 meq/100g soils

(*b*) 250–1500 meq/100g soils

(*c*) 100–1500 meq/100g soils

(*d*) 100–150 meq/100g soils

Q. 86 Ion exchange phenomenon was discovered by:

(*a*) Thompson and Way (*b*) Schofield

(*c*) Adams and Holmes (*d*) Weighner

Q. 87 What is unit of cation exchange capacity?

(*a*) meq/100g (*b*) cmol(p+)/kg

(*c*) mmho/100g (*d*) All

Q. 88 What is the SI unit of cation exchange capacity?

(*a*) meq/100g (*b*) cmol(p+)/kg

(*c*) mmho/100g (*d*) All

Q. 89 Equivalency, reversibility and stoichiometry are the characteristics of which of the following?

(*a*) Anion exchange capacity

(*b*) Cation exchange capacity

(*c*) Both (*a*) and (*b*)

(*d*) None

Q. 90 Polymerization of phenols in the presence of formaldehyde will yield which of the following?

(*a*) Anion exchange resin

(*b*) Cation exchange resin

(*c*) Both (*a*) and (*b*)

(*d*) None

Q. 91 Polymerization of p-phenylenediamines in the presence of formaldehyde will yield which of the following?

(*a*) Anion exchange resin

(*b*) Cation exchange resin

(*c*) Both (*a*) and (*b*)

(*d*) None

Q. 92 Ion exchange reactions can be used for:

(*a*) Separation of ions

(*b*) Deionization of water

(*c*) Ion activity measurements

(*d*) All

Q. 93 Softening of water can be done by the use of:

(*a*) Calcium saturated exchange resins

(*b*) Magnesium saturated exchange resins

(*c*) Sodium saturated exchange resins

(*d*) All the above

Q. 94 What are the elements responsible for the hardness of water?

(*a*) N and K (*b*) Ca and Mg

(*c*) Na and K (*d*) Ca, Mg and Na

Q. 95 Permutit is a synthetic ____________, which has cation exchange properties:

(*a*) Aluminosilicates (*b*) Ca- exchange resins

(*c*) Mg- exchange resins (*d*) Na- exchange resins

Q. 96 $X/m = KC^n$ is:

(*a*) Langmuir's equation

(*b*) Freundlich's equation

(*c*) Gapon's equation

(*d*) Vanselow's equation

Q. 97 $X/m = K_1K_2C/1+K_2C$ is:

(*a*) Langmuir's equation

(*b*) Freundlich's equation

(*c*) Gapon's equation

(*d*) Vanselow's equation

Q. 98 What is/are the equation(s) used for homovalent exchange reactions?

(*a*) Gapon's

(*b*) Krishnamoorthy, Davis and Overstreet's

(*c*) Vanselow's

(*d*) Kerr's

(*e*) All the above

Q. 99 What is/are the equation(s) used for heterovalent exchange reactions?

(*a*) Gapon's

(*b*) Vanselow's

(*b*) Krishnamoorthy, Davis and Overstreet's

(*d*) All the above

Q. 100 "Nearest-neighbour" model was proposed by:

(*a*) Vanselow

(*b*) Krishnamoorthy, Davis and Overstreet

(*c*) Gapon

(*d*) Kerr

Q. 101 Langmuir adsorption equation gives:

(*a*) Hyperbolic curve (*b*) Parabolic curve

(*c*) Linear curve (*d*) All

Q. 102 The capability of soil to produce a plant or plant parts is called as:

(*a*) Soil fertility (*b*) Soil quality

(*c*) Soil productivity (*d*) All

Q. 103 The inherent capacity of soil to provide nutrients is called as:

(*a*) Soil fertility (*b*) Soil quality

(*c*) Soil productivity (*d*) All

Q. 104 The total number of essential nutrient elements required for plant growth is:

(*a*) 16 (*b*) 17

(*c*) 20 (*d*) 21

Q. 105 The 'Functional or metabolism nutrients' was given by:

(*a*) Arnon and Stout (*b*) Nicholas

(*c*) Mayor and Brown (*d*) Juston von Liebig

Q. 106 N, P and K are called as:

(*a*) Macro nutrients (*b*) Micronutrients

(*c*) Secondary nutrients (*d*) Beneficial nutrients

Q. 107 Ca, Mg and S are called as:

(*a*) Macro nutrients (*b*) Micronutrients

(*c*) Secondary nutrients (*d*) Beneficial nutrients

Q. 108 Iron, zinc, boron and molybdenum are called as:

(*a*) Macro nutrients (*b*) Micronutrients

(*c*) Secondary nutrients (*d*) Beneficial nutrients

Q. 109 Na, Ve, Ni, Si, Co, Se and I are the examples of:

(*a*) Macro nutrients (*b*) Micronutrients

(*c*) Secondary nutrients (*d*) Beneficial nutrients

Q. 110 The form(s) of phosphorous absorbed by plants is/are:

(*a*) HPO_4^{3-} (*b*) $H_2PO_4^-$

(*c*) HPO_4^{2-} (*d*) Both (*a*) and (*b*)

(*e*) All the above

Q. 111 The major portion of available boron is absorbed by plants as:

(*a*) BO_3^{3-} (*b*) $HB_4O_7^-$

(*c*) $H_2B_4O_7^-$ (*d*) Both (*a*) and (*b*)

Q. 112 Of the following, which is the available form of molybdenum to plant?

(*a*) $HMoO_4^-$ (*b*) MoO_4^-

(*c*) $HMoO_4^{2-}$ (*d*) MoO_4^{2-}

Q. 113 Of the following, which nutrient element(s) is/are essential for protein synthesis?

(*a*) Nitrogen (*b*) Phosphorous

(*c*) Potassium (*d*) Ca and Mg

Q. 114 ____________ is a constituent of sugar phosphates, *viz* ADP, ATP?

(*a*) Nitrogen (*b*) Phosphorous

(*c*) Sulphur (*d*) Molybdenum

Q. 115 Which nutrient element is involved in energy transformation?

(*a*) Molybdenum (*b*) Zinc

(*c*) Sulphate (*d*) Phosphorous

Q. 116 The nutrient element(s) involved in stomatal regulation of cell is/are:

(*a*) Ca and Mg (*b*) Nitrogen

(*c*) Potassium (*d*) Both (*a*) and (*c*)

Q. 117 The nutrient element(s) essential for maintenance of the stability of cell wall is/are:

(*a*) Phosphorous (*b*) Calcium

(*c*) Magnesium (*d*) Potassium

Q. 118 Oil content in oil-bearing plants is increased by:

(*a*) Sulphur (*b*) Molybdenum

(*c*) Nitrogen (*d*) Phosphorous

Q. 119 Of the following, which nutrient element(s) is/are responsible for the translocation of sugars across the membrane?

(*a*) Boron

(*b*) Phosphorus and Zinc

(*c*) Molybdenum and Iron

(*d*) All

Q. 120 Nutrient element(s) essential for photosynthesis is/are:

(*a*) Manganese (*b*) Copper and Zinc

(*c*) Phosphorus (*d*) All

Q. 121 Nutrient element(s) essential for the functioning of sulphydryl compounds such as cysteine is/are:

(*a*) Sulphur (*b*) Zinc

(*c*) Phosphorous (*d*) Both (*a*) and (*b*)

Q. 122 What is the working wavelength range for Fe in atomic absorption spectrophotometer?

(*a*) 0.2-2.0 ppm (*b*) 2-20 ppm

(*c*) 5-50 ppm (*d*) 50-100 ppm

Q. 123 What is the working wavelength range for Mn in atomic absorption spectrophotometer?

(*a*) 0.2-2.0 ppm (*b*) 2-20 ppm

(*c*) 0.5-5 ppm (*d*) 20-50 ppm

Q. 124 What is the working wavelength range for B in atomic absorption spectrophotometer?

(*a*) 2-20 ppm (*b*) 400-1000 ppm

(*c*) 2-5 ppm (*d*) 20-50

Q. 125 What is the working wavelength range for Mo in atomic absorption spectrophotometer?

(*a*) 2.0-20 ppm (*b*) 20-60 ppm

(*c*) 400-1600 ppm (*d*) 0.2-20 ppm

Q. 126 The science dealing with determination of composition of the constituents in terms of element and compounds is called as:

(*a*) Chemical chemistry (*b*) Organic chemistry

(*c*) Analytical chemistry (*d*) Physical chemistry

Q. 127 Law of mass action theory was given by:

(*a*) Arnold (*b*) Guldeberg and Waag

(*c*) Arnon and Stout (*d*) None

Q. 128 Example(s) for the methods of separation is/are:

(*a*) Distillation (*b*) Precipitation

(*c*) Dialysis (*d*) Chromatography

(*e*) All the above

Q. 129 Which of the following is/are responsible for rendering monochromatic light in flame photometer?

(*a*) Collimating mirror (*b*) Burner

(*c*) Atomizer (*d*) All

Q. 130 The main limitation of flame photometer is:

(*a*) Low energy (*b*) Low accuracy

(*c*) Both (*a*) and (*b*) (*d*) None

Q. 131 Atomizer role in flame photometer is to:

(*a*) Introduce the liquid into flame

(*b*) Bring flame at a stable form

(*c*) Bring flame into reproducible rate

(*d*) All the above

Q. 132 The optical system of flame photometer makes the light:

(*a*) Monochromatic (*b*) Multichromatic

(*c*) Both (*a*) and (*b*) (*d*) None

Q. 133 Which of the following is/are measurable below 1 ppm?

(*a*) Ca (*b*) Mg

(*c*) Na and K (*d*) All

Q. 134 Which of the following is measurable below 1-10 ppm?

(*a*) Al and B (*b*) Cd and Co

(*c*) Both (*a*) and (*b*) (*d*) Zn, Au and Bi

Q. 135 Which of the following is measurable below >10 ppm?

(*a*) Zn, Bi and Au (*b*) Fe, Mn, Zn and Cu

(*c*) Mo and B (*d*) All

Q. 136 EDL sensitivity is higher than hallow cathcde lamp because

(*a*) Lower detection limit

(*b*) Higher detection limit

(*c*) Both (*a*) and (*b*)

(*d*) None

Q. 137 Chromatographic separations are carried out by which property (ies) of the substances?

(*a*) Solubility (*b*) Adsorption

(*c*) Volatility (*d*) All

Q. 138 In every chromatographic separation, which of the following is/are the phases?

(*a*) Stationary (*b*) Mobile

(*c*) Intermediate (*d*) Both (*a*) and (*b*)

Q. 139 Partition chromatography deals with

(*a*) Moving liquid phase and stationary phase liquid on support

(*b*) Both the solid phases

(*c*) Both the gaseous phases

(*d*) All the above

Q. 140 Adsorption chromatography deals with

(*a*) Moving liquid phase and stationary phase solid on support

(*b*) Both the solid phases

(*c*) Both the gaseous phases

(*d*) All the above

Q. 141 Suitability of instrumental measurements depend on:

(*a*) Nature of analysis (*b*) Type of heavy metals
(*c*) Level of determination (*d*) Nature of samples
(*e*) All the above

Q. 142 The number of atoms in a flame in a particular excited energy state (*n* *) and the number of atoms in the ground or unexcited state (*n*) are related by the following familiar Boltzman relationship:

(*a*) $n^*/n = A\,\exp(-DE/KT)$
(*b*) $n^*/n = A\,\exp(-DEK/T)$
(*c*) $n^*/n = A\,\exp(-DExKT)$
(*d*) 4. $n^*/n = AK\,\exp(-DE/T)$

Q. 143 n^*/n values for most elements in flames of about 3,000°K are in the range of:

(*a*) 10^{-10} to 10^{-4} range (*b*) $10^{-5}$to 10^{-4} range
(*c*) 10^{-10} to 10^{-5} range (*d*) 10^{-1} to 10^{-2} range

Q. 144 The components of Atomic Absorption Spectrophotometer are:

(*a*) A light source that emits resonance line radiation
(*b*) A sampling cell, usually a flame, into which the sample is fed as an aerosol
(*c*) A monochromator that is used to isolate the absorbing resonance line from other non-absorbing lines
(*d*) All the above

Q. 145 The components of Atomic Absorption Spectrophotometer Instrument are:

(*a*) A detector that measures the amount of absorption
(*b*) Electronic amplification of the absorption signal
(*c*) A readout system that normally is a strip chart recorder, a digital display, a meter, or a printer
(*d*) All the above

Q. 146 Principle of operation in Atomic Absorption Spectrophotometer instruments is:

(*a*) The sample aerosol is aspirated

(*b*) Mixed with the gases before combustion

(*c*) During combustion, the element is reduced to the atomic state and then can absorb radiation

(*d*) Absorption of radiation is directly proportional to no. of atoms present in ground state

(*e*) All the above

Q. 147 Types of Atomic Absorption Instrument are:

(*a*) Single-beam

(*b*) Double- beam

(*c*) Dual-channel double-beam

(*d*) Double-beam sequential multielement

(*e*) All the above

Q. 148 Double beam Atomic Absorption Spectrophotometer instrument compensates for which of the following features?

(*a*) Changes in lamp intensity

(*b*) Variations in electronic noise

(*c*) Detector sensitivity

(*d*) All the above

Q. 149 Non-flame Atomic absorption spectrophotometer is very useful when:

(*a*) Sample size is very restricted

(*b*) Lowest detection limit is needed

(*c*) Both (*a*) and (*b*)

(*d*) None of the above

Q. 150 Which of the following is the principle of Atomic Absorption Spectrophotometer-Graphite Furnace?

(*a*) Pipetting the sample into the HGA

(*b*) Sample is dried at 110°C to remove the water.

(*c*) The temperature is raised to some value at which no atomization of the element in question occurs but where as much as possible of organics or salts are volatized

(*d*) The temperature is raised sufficiently to atomize the element of interest. Typically this is between 2,100 and 2,900°C.

(*e*) All the above

Q. 151 What is/are the advantages of ICP-AES?

(*a*) Multielement analysis capability

(*b*) Large dynamic linear range

(*c*) Reduction of matrix interferences

(*d*) Improved detection limits for refractory elements and

(*e*) Enhancement of productivity

(*f*) All the above

Q. 152 Which of the following is/are the principles of ICP-OES?

(*a*) Sample constituents are vapourized → atomized → excitation of atoms or ions.

(*b*) The excited atoms or ions will emit their specific line spectra.

(*c*) The intensities of these lines depend on the concentrations of their respective elements in the sample

(*d*) Emitted spectra are analyzed by suitable detector

(*e*) All the above

Q. 153 Major components of ICP-OES are:

(*a*) Sample introduction system and Nebulizer

(*b*) Plasma generation assembly

(*c*) RF generator and Detector

(*d*) Both (*a*) and (*c*)

(*e*) All the above

Q. 154 Method of measurement of ICP-OES is:

(*a*) Simultaneous (*b*) Sequential

(*c*) Both (*a*) and (*b*) (*d*) None of the above

Q. 155 Viewing of plasma of ICP-OES is:

(*a*) Axial view (*b*) Radial view

(*c*) Both (*a*) and (*b*) (*d*) None of the above

Q. 156 Detector of ICP-OES is

(*a*) Photomultiplier tube (*b*) Solid state detector

(*c*) Both (*a*) and (*b*) (*d*) None of the above

Q. 157 Which of the following facts is/are true in comparison of AAS with ICP-AES?

(*a*) Furnace AA detection limits are generally better

(*b*) Detection limits for Group I elements (*e.g.* Na, K) are generally better by flame AAS than by ICP

(*c*) Detection limits for refractory elements (*e.g.* B, Ti, V, Al) are better by ICP than by flame AAS

(*d*) Non metals such as phosphorus, sulfur, nitrogen, carbon, and the halogens (*e.g.* I, Cl, Br) can only be determined by ICP

(*e*) All of the above

Q. 158 Which of the following facts is/are true in comparison of Atomic Absorption Spectrophotometer with ICP-AES?

(*a*) Sequential ICP becomes faster than AAS

(*b*) Approximately 6 elements/sample for routine analysis is possible in ICP

(*c*) Both (*a*) and (*b*)

(*d*) None of the above

Q. 159 Which of the following facts is/are true in comparison of Atomic Absorption Spectrophotometer with ICP-AES?

(*a*) Flame AAS cannot be left completely unattended for safety reasons

(*b*) An ICP-ES instrument or graphite furnace AA can be left to run overnight as no combustible gases are involved

(*c*) Both (*a*) and (*b*)

(*d*) None of the above

Q. 160 Which of the following facts is/are true in comparison of Atomic Absorption Spectrophotometer with ICP-AES?

(*a*) In ICP: Cu can be measured from 0.002 ppm to over 200 ppm

(*b*) In AAS: Cu can be measured from 0.02 ppm to over 20 ppm

(*c*) Both (*a*) and (*b*)

(*d*) None of the above

Q. 161 Which of the following facts is/are true in comparison of Atomic Absorption Spectrophotometer with ICP-AES?

(*a*) In AAS, Chemical interferences are high

(*b*) ICP-OES is almost free from chemical interferences due to the high temperatures reached in a plasma

(*c*) Both (*a*) and (*b*)

(*d*) None of the above

Q. 162 Which of the following facts is/are true in comparison of Atomic Absorption Spectrophotometer with ICP-AES?

(*a*) Minimum Spectral interferences in AAS

(*b*) In ICP-OES, greater possibility of spectral interference

(*c*) In sequential ICP, can be overcome by selecting a different wavelength with similar detection limits

(*d*) All the above

Q. 163 Which of the following facts is/are true in comparison of Atomic Absorption Spectrophotometer with ICP-AES?

(*a*) Recalibration is required in case of long duration analysis in case of AAS

(*b*) In ICP, long-term precision is quite high

(*c*) Both (*a*) and (*b*)

(*d*) None of the above

Q. 164 Anodes used in Anodic Stripping Voltammetry is:

(*a*) Indicator electrode

(*b*) Reference electrode (Ag/AgCl/KCl)

(*c*) Auxiliary electrode (Pt)

(*d*) All the above

Q. 165 What is/are the disadvantages in Anodic Stripping Voltammetry?

(*a*) Very slow determination

(*b*) Chemical conditions are very important

(*c*) Sometimes chemical conditions do not permit determination

(*d*) Metals must be in the ionic condition

(*e*) All the above

Q. 166 High Pressure Liquid Chromatography (HPLC) involves which of the following steps

(*a*) Separation of metal species

(*b*) Formation colored complex

(*c*) Determination through absorbance detector

(*d*) All the above

Q. 167 What is/are the basic types of HPLC instrument systems which are available?

(*a*) Isocratic

(*b*) Gradient

(*c*) Both (*a*) and (*b*)

(*d*) None of the above

Q. 168 Which of the following is/are the type of detector being used in HPLC?

(*a*) Absorbance detector

(*b*) Electrochemical detector

(*c*) Conductivity detector

(*d*) All the above

Q. 169 What is/are the components required in HPLC?

(*a*) A gradient/isocratic pump

(*b*) A pre-column CG5A (Dionex)

(*c*) An analytical column CS5A (Dionex)

(*d*) A UV-VIS detector (530 nm)

(*e*) All the above

Q. 170 The type of instrument in a laboratory is decided by which of the following factors?

(*a*) Number of samples/week

(*b*) Type of sample (Water/effluent, soil, compost, etc.)

(*c*) How many elements for each sample

(*d*) What elements

(*e*) All the above

Q. 171 The type of instrument in a laboratory is decided by which of the following factors?

(*a*) What are typical sample volume

(*b*) What concentration range for the element in the samples

(*c*) What expertise is required for the operator

(*d*) Cost of the Instrument

(*e*) Running cost

(*f*) All the above

Q. 172 General formula for quartz is:

(*a*) $R_2 (Si_2O_6)$ (*b*) $RSiO_3$

(*c*) R_2SiO_4 (*d*) SiO_2

Q. 173 What is the general formula for amphiboles?

(*a*) $R_2 (Si_2O_6)$ (*b*) $RSiO_3$

(*c*) R_2SiO_4 (*d*) SiO_2

Q. 174 Which is the most abundant amphibole found in soils?

(*a*) Hypersthene (*b*) Augite

(*c*) Horneblende (*d*) Diapside

Q. 175 Augite is a ____________ group mineral:

(*a*) Quartz (*b*) Pyroxenes

(*c*) Amphiboles (*d*) Olivine

Q. 176 Foresterite and Fayalite are the examples for ________ group minerals:

(*a*) Quartz (*b*) Pyroxenes

(*c*) Amphiboles (*d*) Olivine

Q. 177 Of the following, which is dark coloured mica?

(*a*) Illite (*b*) Muscovite

(*c*) Montmorillonite (*d*) Biotite

Q. 178 Of the following, which is light coloured mica?

(*a*) Illite (*b*) Muscovite

(*c*) Biotite (*d*) Montmorillonite

Q. 179 Apatite is a ________ mineral:

(*a*) Primary (*b*) Secondary

(*c*) Tertiary (*d*) Accessory

Q. 180 What is the percentage of SiO_2 in quartz?

(*a*) 80 (*b*) 90

(*c*) 100 (*d*) 95

Q. 181 Of the following, which one is the example for phosphate mineral?

(*a*) Anorthite (*b*) Apatite

(*c*) Augite (*d*) Biotite

Q. 182 Of the following, which one is the magnesium containing mineral?

(*a*) Hornblende (*b*) Olivine

(*c*) Muscovite (*d*) Albite

Q. 183 Of the following, which one is the high calcium containing mineral?

(*a*) Apatite (*b*) Albite

(*c*) Anorthite (*d*) Augite

Q. 184 Orthoclase minerals can be identified by:

(*a*) Zoning (*b*) Cleavage

(*c*) Partial alteration (*d*) Inclusion

Q. 185 Plagioclase minerals can be identified by:

(*a*) Partial alteration (*b*) Zoning
(*c*) Cleavage (*d*) Inclusion

Q. 186 Mica and gypsum can be identified by:

(*a*) Partial alteration (*b*) Zoning
(*c*) Cleavage (*d*) Inclusion

Q. 187 Plutonic igneous rock that is deficient in silica is:

(*a*) Andesite (*b*) Diorite
(*c*) Granodiorite (*d*) Nepheline

Q. 188 Albite and anorthite can be identified by:

(*a*) Cleavage (*b*) Inclusion
(*c*) Zoning (*d*) Partial alteration

Q. 189 In reflected light, which mineral shows black coloured surface?

(*a*) Apatite (*b*) Magnetite
(*c*) Haematite (*d*) Muscovite

Q. 190 In reflected light, which mineral shows silvery appearance?

(*a*) Ilmenite (*b*) Muscovite
(*c*) Olivine (*d*) Both (*a*) and (*b*)

Q. 191 Pleochroism is a characteristic feature of:

(*a*) Biotite (*b*) Hornblende
(*c*) Tourmaline (*d*) All

Q. 192 Which mineral(s) can be identified by the angle of extinction?

(*a*) Apatite (*b*) Muscovite
(*c*) Augite (*d*) Both (*a*) and (*b*)
(*e*) All the above

Q. 193 Orthoclase feldspar can be distinguished from the quartz and plagioclase feldspar by:

(*a*) Zoning (*b*) Twinning
(*c*) Angle of extinction (*d*) Pleochroism

Q. 194 Which of the following is more stable mineral?

(*a*) Zircon (*b*) Tourmaline

(*c*) Quartz (*d*) Magnetite

Q. 195 Which of the following is easily weatherable mineral?

(*a*) Olivine (*b*) Hornblende

(*c*) Augite (*d*) Biotite

Q. 196 Arrange the following minerals based on their stability in increasing order.

(*a*) Albite > Zircon > Magnetite > Muscovite

(*b*) Magnetite > Muscovite > Albite > Zircon

(*c*) Magnetite > Zircon > Albite > Muscovite

(*d*) Albite > Muscovite > Magnetite > Zircon

Q. 197 Arrange the following minerals based on their stability in decreasing order:

(*a*) Biotite > Apatite > Anorthite > Olivine

(*b*) Anorthite > Apatite > Biotite > Olivine

(*c*) Anorthite > Biotite > Apatite > Olivine

(*d*) Biotite > Anorthite > Apatite > Olivine

Q. 198 Which one of the following is the light mineral?

(*a*) Quartz (*b*) Feldspar

(*c*) Mica (*d*) Microcline

(*e*) All the above

Q. 199 Which one of the following is the heavy mineral?

(*a*) Apatite (*b*) Hornblende

(*c*) Tourmaline (*d*) Zircon

(*e*) All the above

Q. 200 Of the following, which one is the secondary mineral?

(*a*) Clay minerals

(*b*) Amorphous silicates

(*c*) Oxides of iron, aluminium and silicon

(*d*) All the above

Q. 201 Formula for tetrahedral unit is:

(a) Si_2O_6 (b) SiO_2
(c) Si_4O_8 (d) SiO_4

Q. 202 When an oxygen is shared by two adjacent tetrahedral unit, the unit negative charge will be:

(a) Increased (b) Decreased
(c) Unaltered (d) All

Q. 203 Dioctahedral structure will be formed when two thirds of octahedral position is filled with __________ ions:

(a) Trivalent (b) Divalent
(c) Monovalent (d) Both (a) and (b)

Q. 204 Trioctahedral structure will be formed when all the octahedral position are filled with __________ ions:

(a) Trivalent (b) Divalent
(c) Monovalent (d) Both (a) and (b)

Q. 205 Which is an amorphous clay mineral?

(a) Kaolinite (b) Smectite
(c) Allophane (d) Chlorite

Q. 206 Of the following, which is/are example(s) for 1:1 dioctahedral clay minerals?

(a) Serpentine (b) Kaolinite
(c) Halloysite (d) Both (a) and (b)
(e) Both (b) and (c)

Q. 207 Of the following, which is/are example(s) for 1:1 trioctahedral clay minerals?

(a) Greenalite (b) Serpentine
(c) Chrysolite (d) All

Q. 208 The most widely occurring 1:1 type clay minerals is:

(a) Kaolinite (b) Halloysite
(c) Serpentine (d) Greenalite

Q. 209 What is the basal spacing of kaolinite?

(*a*) 10.25 Å (*b*) 7.2 Å

(*c*) 10 Å (*d*) 14 Å

Q. 210 What is the basal spacing of halloysite?

(*a*) 10.25Å (*b*) 7.2 Å

(*c*) 10 Å (*d*) 14 Å

Q. 211 The most dominant and important clay mineral identified in vertisols is:

(*a*) Vermiculite (*b*) Montmorillonite

(*c*) Beidellite (*d*) Nontronite

Q. 212 Beidellite and nontronite are the examples of ____________ group minerals:

(*a*) Kaolin (*b*) Smectite

(*c*) Mica (*d*) Chlorite

Q. 213 Crystal units of montmorillonite are held together by

(*a*) Potassium (*b*) Water

(*c*) Sodium (*d*) Brucite

Q. 214 What is the basal spacing of montmorillonite?

(*a*) 10 Å (*b*) 10.25 Å

(*c*) 7.2 Å (*d*) 14 Å

Q. 215 Of the following, which one is ferric end member of the smectite group series?

(*a*) Montmorillonite (*b*) Beidellite

(*c*) Nontronite (*d*) Vermiculite

Q. 216 Of the following, which one is aluminium end member of the smectite group series?

(*a*) Montmorillonite (*b*) Beidellite

(*c*) Nontronite (*d*) Vermiculite

Q. 217 ____________ is an example for expanding group of clay minerals:

(*a*) Kaoline (*b*) Smectite

(*c*) Mica (*d*) Chlorite

Q. 218 ____________ is an example for 2:1 type of non-expanding clay minerals:

(*a*) Mica (*b*) Smectite

(*c*) Chlorine (*d*) Kaoline

Q. 219 ____________ is an example for 2:1 type of non-expanding clay mineral:

(*a*) Montmorillonite (*b*) Vermiculite

(*c*) Beidellite (*d*) Nontronite

Q. 220 Crystal units of mica group minerals are held together by:

(*a*) Water (*b*) Potassium

(*c*) Magnesium (*d*) Both (*a*) and (*b*)

(*e*) All the above

Q. 221 What is the basal spacing of illite or hydrous mica?

(*a*) 7.2 °A (*b*) 14 °A

(*c*) 10 °A (*d*) 10.25 °A

Q. 222 Muscovite is a____________ clay mineral:

(*a*) 2:1 dioctahedral (*b*) 2: 1 trioctahedral

(*c*) 1: 1 trioctahedral (*d*) 2:1:1 dioctahedral

Q. 223 ________ is an example for 2: 1 trioctahedral, non-expanding clay mineral:

(*a*) Glauconite (*b*) Illite

(*c*) Beidellite (*d*) Phlogobite

Q. 224 Of the following, which is/are the example(s) for 2: 1 dioctahedral, non-expanding clay minerals?

(*a*) Muscovite (*b*) Glauconite

(*c*) Illite (*d*) All

Q. 225 The 2: 1 non-expanding clay mineral ____________ exists both in dioctahedral and trioctahedral forms:

(*a*) Chlorite (*b*) Glauconite

(*c*) Vermiculite (*d*) Illite

Q. 226 Crystal units of vermiculite are held together by:

(*a*) Water (*b*) Potassium

(*c*) Magnesium (*d*) Both (*b*) and (*c*)

Q. 227 Which clay mineral has high net negative charge?

(*a*) Montmorillonite (*b*) Vermiculite

(*c*) Illite (*d*) Kaolinite

Q. 228 Chlorite is a ____________ type clay mineral:

(*a*) 2 : 1 : 1 (*b*) 2 : 1

(*c*) 1 : 1 (*d*) Both (*a*) and (*b*)

Q. 229 Crystals units of chlorite are held together by:

(*a*) Water (*b*) Brucite

(*c*) Potassium (*d*) Magnesium

Q. 230 Dominant element present in the octahedral sheets of chlorite is:

(*a*) Potassium (*b*) Sodium

(*c*) Aluminium (*d*) Magnesium

Q. 231 Charge density is high for which of the following clay minerals?

(*a*) Montmorillonite (*b*) Illite

(*c*) Kaolinite (*d*) Chlorite

Q. 232 Charge density is low for which of the following clay minerals?

(*a*) Montmorillonite (*b*) Illite

(*c*) Kaolinite (*d*) Chlorite

Q. 233 Which clay mineral is hexagonal in shape?

(*a*) Kaolinite (*b*) Halloysite

(*c*) Montmorillonite (*d*) Vermiculite

Q. 234 Elongated tubles is the shape of __________ mineral:

(*a*) Kaolinite (*b*) Halloysite

(*c*) Montmorillonite (*d*) Vermiculite

Q. 235 Irregular flakes are the shape of __________ mineral:

(*a*) Kaolinite (*b*) Halloysite

(*c*) Montmorillonite (*d*) Vermiculite

Q. 236 What is the SiO_2/Al_2O_3 ratio of kaolinite?

(*a*) Less than 2 (*b*) Greater than 2

(*c*) Equal to 2 (*d*) None

Q. 237 What is the SiO_2/Al_2O_3 ratio of montmorillonite?

(*a*) Less than 2 (*b*) Greater than 2

(*c*) Equal to 2 (*d*) None

Q. 238 What is/are the chemical(s) used for the determination of surface area of clay minerals?

(*a*) Ethylene glycol (*b*) EG monoethyl ether

(*c*) Aluminium oxide (*d*) Both (*a*) and (*b*)

(*e*) All the above

Q. 239 Mineral which has more internal surface area is:

(*a*) Vermiculite (*b*) Illite

(*c*) Montmorillonite (*d*) Kaolinite

Q. 240 Mineral which has more external surface area is:

(*a*) Vermiculite (*b*) Illite

(*c*) Montmorillonite (*d*) Kaolinite

Q. 241 What is the total surface area of montmorillonite ($m^2\ kg^{-1}$)?

(*a*) $580–750 \times 10^3$ (*b*) $130–180 \times 10^3$

(*c*) $780–900 \times 10^3$ (*d*) $120–170 \times 10^3$

Q. 242 What is the external surface area of montmorillonite ($m^2\ kg^{-1}$)?

(*a*) $50–70 \times 10^3$ (*b*) $70–100 \times 10^3$

(*c*) $80–100 \times 10^3$ (*d*) $80–150 \times 10^3$

Q. 243 Total surface area of chlorite ($m^2\ kg^{-1}$) is

(*a*) $37–45 \times 10^3$ (*b*) $130–180 \times 10^3$

(*c*) $120–170 \times 10^3$ (*d*) $580–750 \times 10^3$

Q. 244 CEC of montmorillonite clay mineral (cmol (p+) kg^{-1}) is

(*a*) 69–81 (*b*) 57–64

(*c*) 80–150 (*d*) 30–40

Q. 245 What is the CEC of kaolinite (cmol (p+) kg^{-1})?

(*a*) 30–40 (*b*) 3–15

(*c*) 18–22 (*d*) 57–64

Q. 246 Clay minerals can be identified by:

(*a*) X-ray Diffraction Analysis

(*b*) Electron Diffraction Analysis

(*c*) Differential Thermal Analysis

(*d*) All

Q. 247 What Inert material is used in Differential Thermal Analysis for clay minerals identification?

(*a*) Ceramic oxide (*b*) Hydrogen peroxide

(*c*) Aluminium oxide (*d*) Ethylene glycol

Q. 248 Crystallization of clay minerals is a __________ process:

(*a*) Exothermic (*b*) Endothermic

(*c*) Both (*a*) and (*b*) (*d*) Inert

Q. 249 What is the differential thermal peak of montmorillonite?

(*a*) 850–900 °C (*b*) 500–550 °C

(*c*) 400–525 °C (*d*) 480 °C–530 °C

Q. 250 Particles less than __________ cannot be resolved under ordinary microscope:

(*a*) 0.2 μ (*b*) 0.02μ

(*c*) 2.0μ (*d*) 20μ

Q. 251 K_2O content in illite clay mineral is:

(*a*) 5–7 % (*b*) 8–10 %

(*c*) 10–15 % (*d*) 6–10%

ANSWERS

1	(*b*)	23	(*b*)	45	(*b*)	67	(*c*)
2	(*b*)	24	(*d*)	46	(*b*)	68	(*d*)
3	(*d*)	25	(*c*)	47	(*a*)	69	(*d*)
4	(*c*)	26	(*a*)	48	(*c*)	70	(*b*)
5	(*a*)	27	(*b*)	49	(*b*)	71	(*a*)
6	(*b*)	28	(*d*)	50	(*a*)	72	(*d*)
7	(*c*)	29	(*a*)	51	(*b*)	73	(*b*)
8	(*b*)	30	(*c*)	52	(*b*)	74	(*a*)
9	(*d*)	31	(*c*)	53	(*b*)	75	(*c*)
10	(*a*)	32	(*b*)	54	(*b*)	76	(*b*)
11	(*b*)	33	(*c*)	55	(*c*)	77	(*a*)
12	(*a*)	34	(*a*)	56	(*c*)	78	(*c*)
13	(*c*)	35	(*a*)	57	(*c*)	79	(*c*)
14	(*d*)	36	(*c*)	58	(*b*)	80	(*a*)
15	(*b*)	37	(*a*)	59	(*a*)	81	(*b*)
16	(*a*)	38	(*c*)	60	(*a*)	82	(*a*)
17	(*d*)	39	(*a*)	61	(*b*)	83	(*d*)
18	(*d*)	40	(*a*)	62	(*d*)	84	(*c*)
19	(*b*)	41	(*f*)	63	(*a*)	85	(*b*)
20	(*c*)	42	(*d*)	64	(*b*)	86	(*a*)
21	(*a*)	43	(*c*)	65	(*a*)	87	(*d*)
22	(*d*)	44	(*d*)	66	(*b*)	88	(*b*)

89	(*c*)	119	(*a*)	149	(*c*)	179	(*b*)
90	(*b*)	120	(*a*)	150	(*e*)	180	(*c*)
91	(*a*)	121	(*b*)	151	(*f*)	181	(*b*)
92	(*d*)	122	(*b*)	152	(*e*)	182	(*b*)
93	(*c*)	123	(*c*)	153	(*e*)	183	(*c*)
94	(*b*)	124	(*b*)	154	(*c*)	184	(*c*)
95	(*a*)	125	(*b*)	155	(*c*)	185	(*b*)
96	(*b*)	126	(*c*)	156	(*c*)	186	(*c*)
97	(*a*)	127	(*b*)	157	(*e*)	187	(*d*)
98	(*d*)	128	(*e*)	158	(*c*)	188	(*c*)
99	(*d*)	129	(*a*)	159	(*c*)	189	(*b*)
100	(*b*)	130	(*a*)	160	(*c*)	190	(*a*)
101	(*a*)	131	(*d*)	161	(*c*)	191	(*d*)
102	(*c*)	132	(*a*)	162	(*d*)	192	(*e*)
103	(*a*)	133	(*d*)	163	(*c*)	193	(*b*)
104	(*b*)	134	(*c*)	164	(*d*)	194	(*a*)
105	(*b*)	135	(*a*)	165	(*e*)	195	(*a*)
106	(*a*)	136	(*a*)	166	(*d*)	196	(*d*)
107	(*c*)	137	(*d*)	167	(*c*)	197	(*b*)
108	(*b*)	138	(*d*)	168	(*d*)	198	(*e*)
109	(*d*)	139	(*a*)	169	(*e*)	199	(*e*)
110	(*d*)	140	(*a*)	170	(*e*)	200	(*d*)
111	(*b*)	141	(*e*)	171	(*f*)	201	(*d*)
112	(*a*)	142	(*a*)	172	(*d*)	202	(*b*)
113	(*a*)	143	(*a*)	173	(*b*)	203	(*a*)
114	(*b*)	144	(*d*)	174	(*c*)	204	(*b*)
115	(*d*)	145	(*d*)	175	(*b*)	205	(*c*)
116	(*c*)	146	(*e*)	176	(*d*)	206	(*e*)
117	(*b*)	147	(*e*)	177	(*d*)	207	(*d*)
118	(*a*)	148	(*d*)	178	(*b*)	208	(*a*)

209	**(*b*)**	**220**	**(*e*)**	**231**	**(*b*)**	**242**	**(*d*)**
210	**(*a*)**	**221**	**(*c*)**	**232**	**(*c*)**	**243**	**(*b*)**
211	**(*b*)**	**222**	**(*a*)**	**233**	**(*a*)**	**244**	**(*c*)**
212	**(*b*)**	**223**	**(*d*)**	**234**	**(*b*)**	**245**	**(*b*)**
213	**(*b*)**	**224**	**(*d*)**	**235**	**(*c*)**	**246**	**(*d*)**
214	**(*d*)**	**225**	**(*c*)**	**236**	**(*a*)**	**247**	**(*c*)**
215	**(*c*)**	**226**	**(*c*)**	**237**	**(*b*)**	**248**	**(*b*)**
216	**(*b*)**	**227**	**(*c*)**	**238**	**(*e*)**	**249**	**(*a*)**
217	**(*b*)**	**228**	**(*a*)**	**239**	**(*c*)**	**250**	**(*a*)**
218	**(*a*)**	**229**	**(*b*)**	**240**	**(*a*)**	**251**	**(*d*)**
219	**(*b*)**	**230**	**(*d*)**	**241**	**(*a*)**		

Chapter 9

Soil Physics

Q. 1 Who is the Father of soil physics?

(*a*) Hillel (*b*) Richards

(*c*) Dokuchaiev (*d*) Thomas Way

Q. 2 Organic matter in soil can be decomposed by:

(*a*) Hydrogen peroxide (*b*) Sodium carbonate

(*c*) Sodium hydroxide (*d*) Sulphuric acid

Q. 2 What is the number of classifications in size groups of soil particles by International Society of Soil Science (ISSS)?

(*a*) Three (*b*) Four

(*c*) Six (*d*) Seven

Q. 4 What is the size of silt particles in ISSS classification of soil texture?

(*a*) < 2 µ (*b*) 2–20 µ

(*c*) 20–500µ (*d*) 50 –100 µ

Q. 5 What is the size of fine sand particles in USDA classification of soil texture?

(*a*) 0.02–0.1µ (*b*) 0.05–0.1µ

(*c*) 0.1–0.25µ (*d*) 0.02–0.25µ

Q. 6 What is the size of fine clay particles in USDA classification of soil texture?

(*a*) <2 mm (*b*) <0.2 mm

(*c*) <0.02 mm (*d*) <0.002 mm

Q. 7 Surface area of spheroid can be calculated by which of the following formula?

(*a*) πr^3 (*b*) $4/3\ \pi r^3$

(*c*) $2\ \pi r$ (*d*) $\pi r^2 h$

Q. 8 Particle smaller than _______ microns cannot be separated by sieving:

(*a*) <2μ (*b*) < 20μ

(*c*) < 50μ (*d*) < 100μ

Q. 9 What is the unit of viscosity?

(*a*) cm sec^{-1} (*b*) g cm^{-3}

(*c*) g sec cm^{-1} (*d*) cm sec^{-2}

Q. 10 What is the unit of gravity?

(*a*) cm sec^{-1} (*b*) g cm^{-3}

(*c*) g sec cm^{-1} (*d*) cm sec^{-2}

Q. 11 The equation, $v = K.r^2$ is called as:

(*a*) Stoke's law (*b*) Ratio law

(*c*) Liebig law (*d*) Mitcherlich equation

Q. 12 In Stoke's law 'η' denotes:

(*a*) Density (*b*) Gravity

(*c*) Viscosity (*d*) Velocity

Q. 13 Textural analysis of all soils except calcareous and organic soils can be done by:

(*a*) Hydrometer method

(*b*) International pipette method

(*c*) Aggregate method

(*d*) All

Q. 14 More accurate analysis of soil texture can be done by:

(*a*) Hydrometer method

(*b*) International pipette method

(*c*) Aggregate method

(*d*) All

Q. 15 What is the value of particle density in Stoke's law?

(*a*) 1.1-1.8 g cm^{-3} (*b*) 2.25 g cm

(*c*) 2.45 g cm^{-3} (*d*) 2.65 g cm^{-3}

Q. 16 Soils that have 60 % clay, 20 % sand and 20 % silt can be grouped into the textural class of:

(*a*) Clay (*b*) Silty clay

(*c*) Silt (*d*) Silty clay loam

Q. 17 Plasticity and stickiness are high in:

(*a*) Sandy soils (*b*) Loamy soils

(*c*) Clay soils (*d*) All

Q. 18 At the same water content, water availability is more in which of the following soils?

(*a*) Clay soils (*b*) Sandy soils

(*c*) Loamy soils (*d*) All

Q. 19 In which of the following soils, water infiltration is more rapid?

(*a*) Sandy soils (*b*) Loamy soils

(*c*) Clay soils (*d*) All

Q. 20 Of the following, which is/are the primary soil particles?

(*a*) Clay (*b*) Silt

(*c*) Sand (*d*) All

Q. 21 The arrangement of primary and secondary soil particles is called as:

(*a*) Soil structure (*b*) Soil texture

(*c*) Soil consistency (*d*) Soil plasticity

Q. 22 The relative proportion of soil particles is called as

(*a*) Soil structure (*b*) Soil texture

(*c*) Soil consistency (*d*) Soil plasticity

Q. 23 Wet bulk density (g cm^{-3}):

(a) $\dfrac{\text{Weight of soil}}{\text{Volume of soil}}$

(b) $\dfrac{\text{Weight of dry soil}}{\text{Volume of soil}}$

(c) $\dfrac{\text{Weight of dry soil}}{\text{Volume of soil particles}}$

(d) $\dfrac{\text{Volume of soil}}{\text{Weight of moist soil}}$

Q. 24 Dry bulk density (g cm^{-3}):

(a) $\dfrac{\text{Weight of soil}}{\text{Volume of soil}}$

(b) $\dfrac{\text{Weight of dry soil}}{\text{Volume of soil}}$

(c) $\dfrac{\text{Weight of dry soil}}{\text{Volume of soil particles}}$

(d) $\dfrac{\text{Volume of soil}}{\text{Weight of moist soil}}$

Q. 25 Particle density (g cm^{-3}):

(a) $\dfrac{\text{Weight of soil}}{\text{Volume of soil}}$

(b) $\dfrac{\text{Weight of dry soil}}{\text{Volume of soil}}$

(c) $\dfrac{\text{Weight of dry soil}}{\text{Volume of soil particles}}$

(d) $\dfrac{\text{Volume of soil}}{\text{Weight of moist soil}}$

Q. 26 What is the bulk density of fine textured soils?

(*a*) 1.1–1.3 g cm^{-3} (*b*) 2.65 g cm^{-3}

(*c*) 1.4–1.8 g cm^{-3} (*d*) 2.4–2.6 g cm^{-3}

Q. 27 What is the bulk density of coarse textured soils?

(*a*) 2.4–2.6 g cm^{-3} (*b*) 2.65 g cm^{-3}

(*c*) 1.1–1.3 g cm^{-3} (*d*) 1.4–1.8 g cm^{-3}

Q. 28 Value of bulk density is:

(*a*) Higher than particle density

(*b*) Lower than particle density

(*c*) Equal to particle density

(*d*) Unity

Q. 29 Percentage of pore space:

(*a*) [1–Bulk density/Particle density] × 100

(*b*) [10–Bulk density/Particle density] × 100

(*c*) [1–Particle density/Bulk density] × 100

(*d*) [10–Particle density/Bulk density] × 100

Q. 30 Volume wetness = Mass wetness × ________

(*a*) Bulk density (*b*) Particle density

(*c*) Weight of soil (*d*) Pore space

Q. 31 What is the depth of plough layer of soil?

(*a*) 0–10 cm (*b*) 0–15 cm

(*c*) 0–20 cm (*d*) 0–25 cm

Q. 32 What is the weight of a hectare of surface soil?

(*a*) 2.25×10^6 kg ha^{-1} (*b*) 2×10^6 kg ha^{-1}

(*c*) 2.4×10^6 kg ha^{-1} (*d*) 2.6×10^6 kg ha^{-1}

Q. 33 Element essential for the flocculation of soil particles is

(*a*) Sodium (*b*) Potassium

(*c*) Calcium (*d*) Magnesium

Q. 34 Element essential for the deflocculation of soil particles is:

(*a*) Potassium (*b*) Calcium

(*c*) Magnesium (*d*) Sodium

Q. 35 Zeta potential of soil particles is reduced by:

(*a*) Potassium (*b*) Calcium

(*c*) Sodium (*d*) Magnesium

Q. 36 Increase in zeta potential of soil particles leads to:

(*a*) Flocculation (*b*) Deflocculation

(*c*) Aggregation (*d*) Both (*a*) and (*b*)

(*e*) Both (*a*) and (*c*)

Q. 37 Cementation of soil particles may be caused by:

(*a*) Cations

(*b*) Oxides of iron and aluminium

(*c*) Anions

(*d*) Humus substances

(*e*) All

Q. 38 Theory of bonding of clay–clay particles was given by:

(*a*) E.J. Russell (*b*) R.D. Jenny

(*c*) Juston von Leibig (*d*) Waag and Guldberg

Q. 39 Anion responsible for soil aggregation is:

(*a*) Nitrate (*b*) Phosphate

(*c*) Sulphate (*d*) Molybdate

Q. 40 Hydroxides of iron and aluminium on ________ may form stable aggregates in laterite soils:

(*a*) Oxidation (*b*) Reduction

(*c*) Hydration (*d*) Dehydration

Q. 41 More stable aggregates are formed by:

(*a*) Humus substances (*b*) Earthworms

(*c*) Plant roots decay (*d*) All

Q. 42 Extent and stability of soil aggregates can be measured by:

(*a*) Direct methods of soil structure evaluation

(*b*) Indirect methods of soil structure evaluation

(*c*) Both (*a*) and (*b*)

(*d*) All the above

Q. 43 Direct methods of soil structure evaluation characterize the:

(*a*) Shape, size and arrangement of soil aggregates

(*b*) Extent and stability of soil aggregates

(*c*) Properties related to soil structure

(*d*) All the above

Q. 44 Direct methods of soil structure characterization and evaluation include:

(*a*) Microscopic methods

(*b*) Macroscopic methods

(*c*) Wet and dry sieving methods

(*d*) Both (*a*) and (*b*)

(*e*) All the above

Q. 45 Indirect methods of soil structure characterization and evaluation include:

(*a*) Wet and dry sieving methods

(*b*) Mean weight diameter and GMD

(*c*) Stability index and structural co-efficient

(*d*) All the above

Q. 46 In puddled soil, the soil particles are:

(*a*) Cohesive (*b*) Massive

(*c*) Non-cohesive (*d*) Both (*a*) and (*b*)

(*e*) All the above

Q. 47 The soil structural classification followed in India is/are:

(*a*) ISSS classification (*b*) USDA classification

(*c*) Atterberg classification (*d*) All

Q. 48 Which soil structure is most desirable for agriculture?

(*a*) Platy (*b*) Blocky

(*c*) Crumb (*d*) Granular

Q. 49 Which one of the following is spherical and porous soil structure?

(*a*) Columnar (*b*) Prismatic

(*c*) Granular (*d*) Crumb

Q. 50 What is the total number of soil structural classes?

(*a*) Four (*b*) Five

(*c*) Six (*d*) Seven

Q. 51 The method of soil structure characterization and evaluation that was described by Chepil is:

(*a*) Stability index

(*b*) Wet sieving

(*c*) Dry sieving

(*d*) Mean weight diameter

Q. 52 The method of soil structure characterization and evaluation that was described by Yodder is:

(*a*) Stability index

(*b*) Wet sieving

(*c*) Dry sieving

(*d*) Mean weight diameter

Q. 53 For better soil structure, the value of cumulative mean weight diameter should be:

(*a*) Lower (*b*) Higher

(*c*) Unity (*d*) Zero

Q. 54 For better soil structure, the value of structural co-efficient should be:

(*a*) Lower (*b*) Higher

(*c*) Unity (*d*) Zero

Q. 55 The physical condition of soil in relation to plant growth is called as:

(*a*) Soil tilth (*b*) Tillage
(*c*) Coherence (*d*) Adherence

Q. 56 Which of the following is an optimum soil condition for the tillage operation to produce aggregates of suitable sizes?

(*a*) Dry (*b*) Plastic
(*c*) Friable (*d*) Sticky

Q. 57 Crops conducive for the formation of well structured soils:

(*a*) Grasses (*b*) Cereals
(*c*) Millets (*d*) Oilseeds

Q. 58 Which of the following fertilizers improve soil structure?

(*a*) Nitrogenous fertilizers (*b*) Phosphatic fertilizers
(*c*) Potassic fertilizers (*d*) Sulphate fertilizers

Q. 59 The process of water entry into soil through the surface may be either downward or lateral or both is called as:

(*a*) Percolation (*b*) Infiltration
(*c*) Seepage (*d*) All

Q. 60 The distance traveled by water through soil column is called as:

(*a*) Infiltration rate (*b*) Percolation rate
(*c*) Seepage rate (*d*) Run-off loss

Q. 61 The most important factor(s) that influences the infiltration of water is/are:

(*a*) Soil structure (*b*) Soil texture
(*c*) Porosity (*d*) All

Q. 62 Infiltration rate is influenced by:

(*a*) Soil texture (*b*) Soil structure
(*c*) Porosity (*d*) All the above
(*e*) Initial soil water content

Q. 63 The value of atmospheric pressure is:

(*a*) 1.013×10^6 dynes cm^{-2} (*b*) 1.013 bars

(*c*) 101.3 kPa (*d*) All

Q. 64 1 bar:

(*a*) 100 kPa (*b*) 106 kPa

(*c*) 103 Pa (*d*) 10^3 dynes cm^{-2}

Q. 65 1 bar or 100 kPa is equal to pressure exerted by:

(*a*) 100 cm of water column

(*b*) 1000 cm of water column

(*c*) 10000 cm of water column

(*d*) 1 cm of water column

Q. 66 In soil water column, pressure head increases linearly from the surface downwards (True/False):

(*a*) True (*b*) False

Q. 67 In soil water column, gravity head increases linearly above the reference level (True/False):

(*a*) True (*b*) False

Q. 68 Total head drop ($\Delta H/L$) per unit distance in the soil water column is called as:

(*a*) Flux

(*b*) Flux density

(*c*) Hydraulic gradient

(*d*) Hydraulic conductivity

Q. 69 Rate of flow of liquid through a porous medium under unit hydraulic gradient is called as:

(*a*) Darcy's law

(*b*) Hydraulic conductivity

(*c*) Flux density

(*d*) Stoke's law

Q. 70 The formula $q = k\Delta H/L$ is:

(*a*) Darcy's law (*b*) Hydraulic gradient

(*c*) Flux density (*d*) Fick's law

Q. 71 Hydraulic conductivity of soil is found to increase in which of the following order?

(*a*) Na–soil > K–soil > Ca–soil

(*b*) Ca–soil > K–soil > Na–soil

(*c*) Ca–soil > Na–soil > K–soil

(*d*) Na–soil > Ca–soil > K–soil

Q. 72 Permeability classification of soil was given by:

(*a*) O 'Neal' (*b*) Atterberg

(*c*) Schofield (*d*) Chepil and Woodruff

Q. 73 What is the total permeability class of soil given by O 'Neal'?

(*a*) Four (*b*) Five

(*c*) Six (*d*) Seven

Q. 74 What is the unit of hydraulic conductivity?

(*a*) cm sec^{-2} (*b*) m sec^{-1}

(*c*) cm sec^{-1} (*d*) m sec^{-2}

Q. 75 In the laboratory, hydraulic conductivity is measured by

(*a*) Piezometer

(*b*) Constant head permeameter

(*c*) Auger hole method

(*d*) Both (*a*) and (*b*)

Q. 76 In the field, hydraulic conductivity is measured by:

(*a*) Piezometer

(*b*) Constant head permeameter

(*c*) Auger hole method

(*d*) Both (*a*) and (*b*)

Q. 77 What is the unit of surface tension?

(*a*) dynes cm^{-1} (*b*) cm sec^{-1}

(*c*) g cm^{-3} (*d*) Poise

Q. 78 What is the hydraulic conductivity of the permeability class 'rapid' given by O 'Neal'?

(*a*) 60–70 × cm sec^{-1} (*b*) 170–650 × cm sec^{-1}

(*c*) 350–700 × cm sec^{-1} (*c*) > 700 × cm sec^{-1}

Q. 79 When soil water potential is at hydrostatic pressure greater than atmospheric pressure, the pressure potential is:

(*a*) Unity (*b*) Positive

(*c*) Negative (*d*) Zero

Q. 80 Capillary rise of water is measured by the formula h = _____.

(*a*) 0.15/r (*b*) 0.25/r

(*c*) 0.35/r (*d*) 0.5/r

Q. 81 The potential results from adsorptive and capillary forces of the soil matrix acting on soil water is called as:

(*a*) Gravitational potential (*b*) Matric potential

(*c*) Osmotic potential (*d*) Solute potential

Q. 82 The potential results form the earth's gravitational force is called as:

(*a*) Gravitational potential (*b*) Matric potential

(*c*) Osmotic potential (*d*) Solute potential

Q. 83 The potential results from the presence of solutes in soil water is called as:

(*a*) Solute potential

(*b*) Osmotic potential

(*c*) Matric potential

(*d*) Gravitational potential

(*e*) Both (*a*) and (*b*)

Q. 84 At all points above the reference level, the gravitational potential will be

(*a*) Unity (*b*) Zero

(*c*) Positive (*d*) Negative

Q. 85 With increase in solute content in soil water, the vapour pressure is:

(*a*) Increased (*b*) Decreased

(*c*) Unaltered (*d*) All

Q. 86 What is the unit of soil water potential on mass basis?

(*a*) Joules kg^{-1} (*b*) Dynes cm^{-2}

(*c*) cm or m (*d*) Newtons

Q. 87 What is the unit of soil water potential on volume basis?

(*a*) Joules kg^{-1} (*b*) Dynes cm^{-2}

(*c*) Newtons (*d*) Both (*b*) and (*c*)

(*e*) All the above

Q. 88 What is the unit of soil water potential on weight basis?

(*a*) Joules kg^{-1} (*b*) Dynes cm^{-2}

(*c*) cm or m (*d*) Newtons

Q. 89 1 bar:

(*a*) 10 KN m^{-2} (*b*) 100 KN m^{-2}

(*c*) 10 Nm^{-2} (*d*) 100 Nm^{-2}

Q. 90 1 Joule:

(*a*) 10^5 ergs (*b*) 10^6 ergs

(*c*) 10^7 ergs (*d*) 10^9 ergs

Q. 91 What is the soil water potential at saturation in soil?

(*a*) –15 bars (*b*) 0 bars

(*c*) –31 bars (*d*) –1 bar

Q. 92 What is the soil water potential at air dryness?

(*a*) –15 bars (*b*) 0 bars

(*c*) –31 bars (*d*) –1 bar

Q. 93 What is the soil water potential at field capacity?

(*a*) 10–33 kPa (*b*) 33 kPa

(*c*) 1500 kPa (*d*) 33–1500 kPa

Q. 94 What is the soil water potential at permanent wilting point?

(*a*) 10–33 kPa (*b*) 33 kPa

(*c*) 1500 kPa (*d*) 33–1500 kPa

Q. 95 What is the soil water potential at plant available water?

(*a*) 10–33 kPa (*b*) 33 kPa

(*c*) 1500 kPa (*d*) 33–1500 kPa

Q. 96 Which instrument is used for the measurement of negative soil water pressure (suction)?

(*a*) Hydrometer (*b*) Piezometer

(*c*) Tensiometer (*d*) Infiltrometer

Q. 97 Tensiometer can be used to measure soil water pressure up to:

(*a*) 700–800 cm (*b*) 800–900 cm

(*c*) 900–1000 cm (*d*) 1500 cm

Q. 98 The pF scale was given by:

(*a*) Sorenson (*b*) Atterberg

(*c*) White and Beckett (*d*) Schofield

Q. 99 All parts of tensiometer are filled with which of the following?

(*a*) Water (*b*) Mercury

(*c*) HCl (*d*) Lead/silver chloride

Q. 100 For the measurement of soil water potential beyond 900 cm by tensiometer, _________ can be use in tensiometer instead of water:

(*a*) HCl (*b*) Mercury

(*c*) Platinum (*d*) AgCl

Q. 101 Soil water potential in field can be measured by:

(*a*) Tensiometer

(*b*) Porous cup apparatus

(*c*) Pressure plate apparatus

(*d*) Both (*b*) and (*c*)

(*e*) All the above

Q. 102 In the laboratory, soil water potential can be measured by:

(*a*) Porous cup apparatus

(*b*) Pressure plate apparatus

(*c*) Tensiometer

(*d*) Both (*a*) and (*b*)

(*e*) All the above

Q. 103 In the laboratory, soil water potential can be measured by porous cup apparatus up to:

(*a*) 10 bars (*b*) 5 bars

(*c*) 1 bar (*d*) 15 bars

Q. 104 In the laboratory, soil water potential can be measured by pressure plate apparatus up to:

(*a*) 15 bars (*b*) 10 bars

(*c*) 5 bars (*d*) 1 bar

Q. 105 In the field, tensiometer can be used to measure soil water suction up to:

(*a*) 0.75 bars (*b*) 0.85 bars

(*c*) 0.95 bars (*d*) 1.0 bar

Q. 106 Water held in the pores and drained by 100 cm of suction is called as:

(*a*) Partially available water (*b*) Available water

(*c*) Drainage water (*d*) Both (*a*) and (*b*)

Q. 107 Percentage of pores which is drained by 100 cm of suction is called as:

(*a*) Aeration porosity (*b*) Porosity

(*c*) Water holding capacity. (*d*) Wilting point

Q. 108 What is the size of the macro or non-capillary pores?

(*a*) >6 μ (*b*) >10 μ

(*c*) >15 μ (*d*) >20 μ

Q. 109 What is the size of micro or capillary pores?

(*a*) >6 μ (*b*) >10 μ

(*c*) >15 μ (*d*) >20 μ

Q. 110 What is the suction of partially available water in soil?

(*a*) 33–1500 kPa (*b*) 10–33 kPa

(*c*) 10–100 kPa (*d*) 33–100 kPa

Q. 111 Arrange the following soil texture in the order of decreasing water retention capacity.

(*a*) Loam > Clay > Sandy

(*b*) Sandy > Loam > Clay

(*c*) Clay > Loam > Sandy

(*d*) Sandy > Clay > Loam

Q. 112 Arrange the following soils in decreasing order of water retention capacity:

(*a*) Laterite > Alluvial > Black

(*b*) Alluvial > Laterite > Black

(*c*) Black > Laterite > Alluvial

(*d*) Black > Alluvial > Laterite

Q. 113 *In-situ* measurement of soil water can be done by:

(*a*) Electrical resistance method

(*b*) Neutron scattering method

(*c*) Gamma-rays attenuation technique

(*d*) All the above

Q. 114 Which of the following blocks is/are used in electrical resistance method of soil water estimation?

(*a*) Gypsum (*b*) Nylon

(*c*) Fibre glass (*d*) All

Q. 115 What is the source of gamma radiation in gamma-ray scanner used for soil water estimation?

(*a*) Cesium 137 (*b*) Strontium

(*c*) Lanthanum (*d*) Silver

Q. 116 What is the percentage of N_2 in atmospheric air?

(*a*) 25.0 (*b*) 79.0

(*c*) 20.97 (*d*) 72.0

Q. 117 What is the percentage of carbon dioxide in atmospheric air?

(*a*) 0.03 (*b*) 0.3

(*c*) 0.003 (*d*) 3

Q. 118 What is the percentage of carbon dioxide in soil air?

(*a*) 0.03 (*b*) 0.003

(*c*) 0.3 (*d*) 3

Q. 119 More than 90 per cent of gaseous exchange in soil takes place by:

(*a*) Mass flow (*b*) Diffusion

(*c*) Partial diffusion (*d*) All

Q. 120 Diffusion of gases under total gas pressure gradient takes place by:

(*a*) Mass flow (*b*) Diffusion

(*c*) Partial diffusion (*d*) All

Q. 121 Heat capacity (on oven dry basis) = Specific heat × ________

(*a*) Bulk density (*b*) Volume

(*c*) Weight (*d*) Particle density

Q. 122 Heat capacity (on volume basis) = Specific heat × ________

(*a*) Bulk density (*b*) Volume

(*c*) Weight (*d*) Particle density

Q. 123 What is the unit of heat capacity on volume basis?

(*a*) cal cm^{-3} (*b*) cal g^{-1}

(*c*) kcal cm^{-3} (*d*) kcal g^{-1}

Q. 124 What is the unit of heat capacity on weight basis?

(*a*) cal cm^{-3} (*b*) cal g^{-1}

(*c*) kcal cm^{-3} (*d*) kcal g^{-1}

Q. 125 What is the unit of specific heat?

(*a*) kcal g^{-1} (*b*) kcal cm^{-3}

(*c*) cal cm^{-3} (*d*) cal g^{-1}

Q. 126 What is the unit of thermal gradient?

(*a*) J kg^{-1} $^{o}C^{-1}$ (*b*) J sec^{-1} $^{o}C^{-1}$

(*c*) cal cm^{-1} sec^{-1} $^{o}C^{-1}$ (*d*) cal sec^{-1} $^{o}C^{-1}$

Q. 127 Thermal conductivity follows the order of:

(*a*) Peat > Clay > Loam > Sand

(*b*) Sand > Loam > Clay > Peat

(*c*) Loam > Peat > Clay > Sand

(*d*) Peat > Loam > Clay > Sand

Q. 128 The heat absorbed by soil during day on sunny days is lost during night by:

(*a*) Radiation (*b*) Conduction

(*c*) Convection (*d*) Transmission

Q. 129 The soil temperature regimes classification followed in India is/are:

(*a*) ISSS classification (*b*) USDA classification

(*c*) Russian classification (*d*) Both (*a*) and (*b*)

Q. 130 The dominant spectral colour in Munsell colour chart indicates:

(*a*) Hue (*b*) Value

(*c*) Chroma (*d*) All

Q. 131 In soil colour determination, Hue is related to:

(*a*) Quality of light (*b*) Intensity of light

(*c*) Wavelength of light (*d*) All

Q. 132 Relative purity or strength of spectral colour in Munsell colour chart indicates:

(*a*) Hue (*b*) Value

(*c*) Chroma (*d*) All

Q. 133 In soil colour determination, Chroma is related to:

(*a*) Quality of light (*b*) Intensity of light

(*c*) Wavelength of light (*d*) All

Q. 134 In soil colour determination, Value is related to:

(*a*) Quality of light (*b*) Intensity of light

(*c*) Wavelength of light (*d*) All

Q. 135 What is the total number of primary colours available in Munsell colour chart?

(*a*) Two (*b*) Three

(*c*) Four (*d*) Five

Q. 136 What is the range of Hue in Munsell colour chart?

(*a*) 0–10 (*b*) 1–10

(*c*) 0–9 (*d*) 1–9

Q. 137 What is the range of Value in Munsell colour chart?

(*a*) 1–10 (*b*) 0–9

(*c*) 0–10 (*d*) 1–9

Q. 138 In Munsell colour chart, Hue number zero denotes:

(*a*) Absolute white (*b*) Absolute block

(*c*) Absolute yellow (*d*) Absolute red

Q. 139 In Munsell colour chart, Hue number 10 denotes:

(*a*) Absolute white (*b*) Absolute block

(*c*) Absolute yellow (*d*) Absolute red

Q. 140 In Munsell colour chart, Value number zero denotes:

(*a*) Absolute white (*b*) Absolute block

(*c*) Absolute yellow (*d*) Absolute red

Q. 141 In Munsell colour chart, Value number 10 denotes:

(*a*) Absolute white (*b*) Absolute block

(*c*) Absolute yellow (*d*) Absolute red

Q. 142 What is the range of Chroma in Munsell colour chart?

(*a*) 0–10 (*b*) 0–8

(*c*) 1–10 (*d*) 0–9

Q. 143 Diffusion of gases under the concentration gradient takes place by:

(*a*) Mass flow (*b*) Diffusion

(*c*) Root interception (*d*) All

Q. 144 Water content at which soil will start to crumble when it is rolled into a thread under the palm of the hand and this limit is called as:

(*a*) Lower limit

(*b*) Liquid limit or upper plastic limit

(*c*) Sticky limit

(*d*) Plastic limit or lower plastic limit

Q. 145 Water content above which a mixture of soil and water flow as a viscous liquid and below which the material is plastic and this limit is called as:

(*a*) Liquid limit or upper plastic limit

(*b*) Sticky limit

(*c*) Plastic limit or lower plastic limit

(*d*) Lower limit or shrinkage limit

Q. 146 Water content above that a mixture of soil and water will adhere to a steel spatula and this limit is called as:

(*a*) Liquid limit or upper plastic limit

(*b*) Plastic limit or lower plastic limit

(*c*) Sticky limit

(*d*) Lower limit or shrinkage limit

Q. 147 Below the shrinkage limit of soil consistency, the soil is:

(*a*) Moist and friable
(*b*) Dry and hard
(*c*) Wet and plastic
(*d*) Moist and plastic

Q. 148 Between the shrinkage limit and the lower plastic limit of soil consistency, the soil is:

(*a*) Moist and friable
(*b*) Dry and hard
(*c*) Wet and plastic
(*d*) Moist and plastic

Q. 149 Between the lower plastic limit and upper plastic limit of soil consistency, the soil is:

(*a*) Moist and friable
(*b*) Dry and hard
(*c*) Wet and plastic
(*d*) Moist and plastic

Q. 150 Difference between the lower plastic limit and upper plastic limit of soil consistency is called as:

(*a*) Sticky limit
(*b*) Shrinkage limit
(*c*) Plastic limit
(*d*) Plasticity index

Q. 151 Swelling and shrinkage properties of soil can be determined by:

(*a*) Co-efficient of linear extensibility (COLE)
(*b*) Potential volume change (PVC)
(*c*) Swell index
(*d*) All the above

Q. 152 Significant swelling and shrinkage of soil is expected when co-efficient of linear extensibility (COLE) exceeds the value of:

(*a*) 0.03
(*b*) 0.06
(*c*) 0.09
(*d*) 1.32

Q. 153 The value of co-efficient of linear extensibility (COLE) higher than ________ indicates the presence of a significant amount of montmorillonite or swelling clay:

(*a*) 0.09
(*b*) 0.03
(*c*) 1.32
(*d*) 0.06

Q. 154 In soil, water pressure head or hydraulic head at saturated condition is measured by:

(*a*) Porus cup apparatus

(*b*) Pressure plate apparatus

(*c*) Piezometer

(*d*) Tensiometer

Q. 155 In soil, water pressure head or hydraulic head under unsaturated condition is measured by:

(*a*) Porus cup apparatus

(*b*) Pressure plate apparatus

(*c*) Piezometer

(*d*) Tensiometer

Q. 156 At high soil moisture suction, water potential is measured by:

(*a*) Porus cup apparatus

(*b*) Pressure plate apparatus

(*c*) Piezometer

(*d*) Tensiometer

Q. 157 At low soil moisture suction, water potential is measured by:

(*a*) Porus cup apparatus

(*b*) Pressure plate apparatus

(*c*) Piezometer

(*d*) Tensiometer

Q. 158 At high water potential, water retention in soil is mainly controlled by:

(*a*) Soil consistency (*b*) Soil porosity

(*c*) Soil structure (*d*) Soil texture

Q. 159 At low water potential, water retention in soil is mainly controlled by:

(*a*) Soil consistency (*b*) Soil porosity

(*c*) Soil structure (*d*) Soil texture

Q. 160 Chemical(s) used as a detector in neutron moisture meter in the estimation of soil moisture:

(*a*) Ba and Ra

(*b*) Americium and Beryllium

(*c*) BF_3

(*d*) Both (*a*) and (*b*)

Q. 161 Which is/are the neutron emitter or source of neutron in neutron moisture meter in the estimation of soil moisture?

(*a*) Ba and Ra

(*b*) Americium and Beryllium

(*c*) BF_3

(*d*) Both (*a*) and (*b*)

Q. 162 What is the apparent dielectric constant for dry state in soil moisture determination?

(*a*) 3–6 (*b*) 10–15

(*c*) 15–20 (*d*) 20–30

Q. 163 What is the apparent dielectric constant for wet state in soil moisture determination?

(*a*) 3–6 (*b*) 10–15

(*c*) 15–20 (*d*) 20–30

Q. 164 Pressure potential/hydrostatic pressure in soil is estimated by:

(*a*) Porus cup apparatus

(*b*) Pressure plate apparatus

(*c*) Piezometer

(*d*) Tensiometer

Q. 165 The equation, $V = \sqrt{2gz}$ is called as:

(*a*) Reynold's equation (*b*) Torricelli's equation

(*c*) Poiseuille's equation (*d*) Furier's equation

Q. 166 In dry condition, gas diffusion is more in coarse textured soil (True/False):

(*a*) True (*b*) False

Q. 167 In wet condition, gas diffusion is more in coarse textured soil (True/False):

(*a*) True (*b*) False

Q. 168 What is the unit of Oxygen Diffusion Rate (ODR)?

(*a*) $g/cm^2/min$ (*b*) cm/g/min

(*c*) min/cm/g (*d*) None

Q. 169 The value of diffusion co-efficient in soil is:

(*a*) 0.2–0.4 (*b*) 0.4–0.6

(*c*) 0.6–0.8 (*d*) 0.8–1.0

Q. 170 The Land Capability Classification (LCC) based on the physical characteristics of soil was developed by:

(*a*) USDA (*b*) Russia

(*c*) ISSS (*d*) India

Q. 171 Bulk density of organic matter is:

(*a*) 0.3 g/cc (*b*) 0.5 g/cc

(*c*) 0.6 g/cc (*d*) 1.0 g/cc

Q. 172 Percentage of organic matter in soil (on volume basis) is:

(*a*) 5 (*b*) 10

(*c*) 25 (*d*) 50

Q. 173 Percentage of organic matter in soil (on weight basis) is:

(*a*) 5 (*b*) 10

(*c*) 25 (*d*) 50

Q. 174 The bulk density of a soil must consider the ______:

(*a*) Total soil volume and oven dry soil weight

(*b*) Volume of soil and volume of pore space

(*c*) Pore volume and weight of soil particles

(*d*) Volume of soil particles and weight of soil particles

Q. 175 The weight of a graduated cylinder is 54 grams, the cylinder is filled with 85 ml of oven dry field soil. The weight of soil and cylinder is 186 grams. The bulk density of the soil is ______ g/cc:

(*a*) 0.65 (*b*) 1.15

(*c*) 1.35 (*d*) 1.55

Q. 176 Diffusion in soil is:

(*a*) The movement of ions down a concentration gradient

(*b*) PO_4 moving with water via gravity flow

(*c*) Movement of ions by electrostatic attraction

(*d*) Evapotranspiration in plants

Q. 177 Matric potential is always negative (True/False):

(*a*) True (*b*) False

Q. 178 During puddling macropores will be more while micropores will be less (True/False):

(*a*) True (*b*) False

Q. 179 Which of the following structure is mostly influenced by the practical methods of soil management?

(*a*) Platy and columnar (*b*) Prism and columnar

(*c*) Granular and crumby (*d*) Prism and platy

Q. 180 Which of the following process is the starting process of aggregate formation?

(*a*) Flocculation (*b*) Deflocculation

(*c*) Compaction (*d*) All

Q. 181 What is the percentage of air in soil below which mostly plant suffer?

(*a*) 10% (*b*) 10-12%

(*c*) 10-15% (*d*) 15-20%

Q. 182 Of the several agroclimatic elements, the broad distribution of crop in India is mainly governed by

(*a*) Rainfall pattern

(*b*) Mean daily temperature

(*c*) Solar energy received

(*d*) Relative humidity

Q. 183 What is the upper limit of available water in soil?

(*a*) Permanent wilting point (*b*) Hygroscopic point

(*c*) Field capacity (*d*) All

Q. 184 Run-off occurs in soil when the infiltration capacity is high (True/False):

(*a*) True (*b*) False

Q. 185 In which of the following soils maximum loss of soil is caused by saltation type of wind erosion?

(*a*) Loamy soils (*b*) Clayey soils

(*c*) Sandy soils (*d*) None

Q. 186 One cusec is equal to how many litres of water?

(*a*) 27.32 (*b*) 28.32

(*c*) 29.32 (*d*) 30.32

Q. 187 Soil surface area ________ as soil particle size ________:

(*a*) Increases; Increases (*b*) Increases; Decreases

(*c*) Decreases; Increases (*d*) Decreases; Decreases

Q. 188 Three types of soil structure includes:

(*a*) Prisms, Blocky, Granular

(*b*) Columnar, Porus, Brittle

(*c*) Fine, Medium and Coarse

(*d*) Brittle, Firm and Friable

(*e*) Sand, Silt and Clay

Q. 189 Stoke's law was given by G.G stokes in:

(*a*) 1851 (*b*) 1861

(*c*) 1871 (*d*) 1891

Q. 190 What is the unit of thermal diffusivity?

(*a*) cm/°C/sec (*b*) cm^2/sec

(*c*) cm/sec^2 (*d*) sec/°C/cm

Q. 191 What is the size of medium pores in soil?

(*a*) < 10 m (*b*) 10–20 m

(*c*) 20–200 m (*d*) 2–20 m

Q. 192 What is the size of fine pores in soil?

(*a*) < 10 m (*b*) 10–20 m

(*c*) 20–200 m (*d*) 2–20 m

Q. 193 What is the unit of Oxygen Diffusion Rate (ODR)?

(*a*) cm/°C/sec (*b*) $g/cm^2/min$

(*c*) sec/°C/cm (*d*) $cm^2/g/min$

Q. 194 What is the critical limit of Oxygen Diffusion Rate?

(*a*) $20 \times 10^{-8}\ g/cm^2/min$ (*b*) 22.9×10^{-8}

(*c*) 3.9×10^{-8} (*d*) 25×10^{-8}

Q. 195 What is the amount of energy required to evaporate one gram of water?

(*a*) 560 cal (*b*) 570 cal

(*c*) 580 cal (*d*) 590 cal

Q. 196 What is the unit of specific heat?

(*a*) Joules/gram (*b*) Joules/kg

(*c*) cal/g (*d*) cal/kg

Q. 197 What is the unit of thermal conductivity?

(*a*) J/°C/cm/sec (*b*) °C/J/cm/sec

(*c*) °C/J/sec/cm (*d*) J/cm/S/°C

Q. 198 What is the unit of viscosity?

(*a*) cm^2/sec (*b*) cm/sec^2

(*c*) Poise (*d*) mg/m^3

Q. 199 Finer is the soil texture, lower will be bulk density (True/False):

(*a*) True (*b*) False

Q. 200 Which soil has more porosity/or pore space?

(*a*) Clayey soils (*b*) Sandy soils

(*c*) Loamy soils (*d*) Sandy loamy soils

Q. 201 With the increase in compactness and wetness of the soils, thermal conductivity will be:

(*a*) Increased (*b*) Decreased

(*c*) No change (*d*) None

Q. 202 Specific heat of soil is:

(*a*) 1 cal/g (*b*) 0.38 cal/g

(*c*) 0.2 cal/g (*d*) 0.44 cal/g

Q. 203 What is the amount of energy released during the decomposition of one mole organic matter?

(*a*) 580 cal/mol (*b*) 478 KJ/mol

(*c*) 478 cal/mol (*d*) 580 KJ/mol

Q. 204 Specific heat of minerals is:

(*a*) 1 cal/g (*b*) 0.44 cal/g

(*c*) 0.11 cal/g (*d*) 0.38 cal/g

Q. 205 What is the specific heat of organic matter?

(*a*) 1 cal/g (*b*) 0.44 cal/g

(*c*) 0.11 cal/g (*d*) 0.38 cal/g

Q. 206 The pF scale was given by:

(*a*) Sorenson (*b*) Schofield

(*c*) Chepil and Woodruff (*d*) White and Beckett

Q. 207 10 cm of water column is equal to the pF scale of:

(*a*) One (*b*) Two

(*c*) Three (*d*) Four

Q. 208 The value of pF scale at saturated condition in soil is:

(*a*) Zero (*b*) 2.53

(*c*) 4.18 (*d*) 7.0

Q. 209 What is the value of pF scale at field capacity?

(a) Zero (b) 2.53

(c) 4.18 (d) 7.0

Q. 210 What is the value of pF scale at oven dry condition?

(a) 2.53 (b) 4.18

(c) 4.50 (d) 7.0

Q. 211 What is the value of pF scale at hygroscopic point?

(a) 2.53 (b) 4.18

(c) 4.50 (d) 7.0

Q. 212 What is the value of pF scale at wilting point?

(a) 4.50 (b) 4.28

(c) 4.18 (d) 4.89

Q. 213 Salt water potential is less than pure water potential.(True or False):

(a) True (b) False

Q. 214 Soils containing __________of colloids exhibits plasticity:

(a) > 5 % (b) > 10 %

(c) > 15 % (d) > 20 %

Q. 215 Arrange the cations based on their flocculating power in decreasing order:

(a) Al > Ca and H > Mg > K > Na

(b) Ca and H > Al > Ms > K > Na.

(c) Ca and H > Mg > Al > K > Na

(d) Na > K > Mg > Ca and H > Al

Q. 216 Soil compaction, increases the:

(a) Bulk density

(b) Hydraulic conductivity

(c) Macrospores

(d) Water intake rate

Q. 217 By soil compaction, which one of the following is decreased?

(*a*) Microspores
(*b*) Thermal conductivity
(*c*) Thermal diffusivity
(*d*) Hydraulic conductivity

Q. 218 Soils that has ________ % of clay prone to puddling:

(*a*) > 10 (*b*) > 20
(*c*) > 25 (*d*) > 30

Q. 219 Puddling increases the aggregated structure in soils (True/ False):

(*a*) True (*b*) False

Q. 220 Volume wetness = Mass wetness x________:

(*a*) Bulk density (*b*) Particle density
(*c*) 1/Bulk density (*d*) 1/Particle density

Q. 221 ________ and ________ are called as the "skeleton" of the soil:

(*a*) Clay.... Silt (*b*) Clay...... Sand
(*c*) Sand...... Silt (*d*) None

Q. 222 Which of the following is called as the "flesh" of the soil?

(*a*) Sand (*b*) Silt
(*c*) Clay (*d*) None

Q. 223 The larger limit of particles exhibiting Brownian movement is approximately ________:

(*a*) 0.002 mm (*b*) 0.0002 mm
(*c*) 0.02 mm (*d*) 0.05 mm

Q. 224 Columnar structure are mostly found in:

(*a*) Salt affected soils (*b*) Sodic soils
(*c*) Black soils (*d*) Both (*a*) and (*b*)
(*e*) All

Q. 225 In the wet sieving technique of soil structures measurement, __________ is more satisfactory as the lower limit of the size of the aggregates:

(*a*) 0.1 mm (*b*) 0.25 mm
(*c*) 0.01 mm (*d*) 0.5 mm

Q. 226 Of the following soils, __________ is occupied by 50 % solid, 25 % air and 25 % water:

(*a*) Silt loam soils (*b*) Loam soils
(*c*) Sandy clay loam soils (*d*) Silt clay loam soils

Q. 227 What is the unit of bulk density/particle density?

(*a*) g/cm^3 (*b*) Mg/m^3
(*c*) Mg/cm^3 (*d*) g/m^3
(*e*) Both (*a*) and (*b*)

Q. 228 The ratio of total volume of soil to the total mass of dry soil is called as:

(*a*) Dry bulk density (*b*) Wet bulk density
(*c*) Specific volume (*d*) Porosity
(*e*) Particle density

Q. 229 Specific volume of soil is expressed as:

(*a*) cm^3/g (*b*) cm^2/g
(*c*) g/cm^3 (*d*) g/cm^2

Q. 230 Specific volume of coarse textured soils varies from:

(*a*) 0.55 to 0.70 cm^3/g (*b*) 0.65 to 0.70 cm^3/g
(*c*) 0.50 to 0.720 cm^3/g (*d*) 0.55 to 0.75 cm^3/g

Q. 231 Specific volume of fine textured soils varies from:

(*a*) 0.75 to 0.95 cm^3/g (*b*) 0.70 to 0.95 cm^3/g
(*c*) 0.75 to 0.90 cm^3/g (*d*) 0.70 to 0.90 cm^3/g

Q. 232 The ratio of total volume of pores spaces to the total volume of soil is called as:

(*a*) Capillary porosity (*b*) Porosity
(*c*) Non-capillary porosity (*d*) Void ratio

Q. 233 The capillary porosity of soil is determined by finding the volume of water retained in the soil at ____________ tension:

(a) 40 cm (b) 45 cm

(c) 50 cm (d) 60 cm

Q. 234 The ratio of total volume of pores to the total volume of soil solids is called as:

(a) Capillary porosity

(b) Non-capillary porosity

(c) Porosity

(d) Void ratio

Q. 235 ____________ is an index of the relative volume of soil pores:

(a) Capillary porosity (b) Porosity

(c) Non-capillary porosity (d) Void ratio

Q. 236 The volume of void ratio in soil usually varies between:

(a) 0.5 and 2.0 (b) 0.3 and 2.0

(c) 1.0 and 2.0 (d) 1.5 and 2.0

Q. 237 The ratio of total volume water occupied in the pore spaces at a specific time to the total volume of soil is called as:

(a) Gravimetric water content (b) Mass wetness

(c) Volume wetness (d) Degree of saturation

Q. 238 The ratio of the volume of water present in the soil at a particular time to the volume of pores is called as:

(a) Gravimetric water content (b) Mass wetness

(c) Volume wetness (d) Degree of saturation

Q. 239 The ratio of volume of air space to the total volume of soil is called as:

(a) Capillary porosity (b) Air-filled porosity

(c) Non-capillary porosity (d) Porosity

Q. 240 ____________ is a measure of the relative air content of the soil:

(*a*) Capillary porosity (*b*) Air-filled porosity
(*c*) Non-capillary porosity (*d*) Porosity

Q. 241 In USDA classification of soil texture, what is the size of medium sand?

(*a*) 0.05–0.10 mm (*b*) 0.10–0.25 mm
(*c*) 0.25–0.5 mm (*d*) 0.5–1.0 mm

Q. 242 In USDA classification of soil texture, what is the size of silt?

(*a*) 0.05–0.10 mm (*b*) 0.10–0.25 mm
(*c*) 0.25–0.5 mm (*d*) 0.002–0.02 mm

Q. 243 Arrange the soil texture based on their textural orders:

(*a*) Sandy loam → Loamy sand → Silty loam → Loam
(*b*) Loamy sand → Sandy loam → Silty loam → Loam
(*c*) Loamy sand → Sandy loam → Loam → Silty loam
(*d*) Sandy loam → Loamy sand → Loam → Silty loam

Q. 244 Soils containing more than ____________ % of organic matter are classified as peats:

(*a*) 20 (*b*) 30
(*c*) 40 (*d*) 50

Q. 245 Soils containing more than ____________ % of organic matter are classified as mucks:

(*a*) 20–30 (*b*) 20–40
(*c*) 20–50 (*d*) 30–50

Q. 246 Clay-to-clay particle interaction, particularly under dry condition, takes place due to:

(*a*) Electrostatic forces (*b*) Cementing agents
(*c*) Van der Waal's forces (*d*) Both (*a*) and (*c*)
(*e*) All the above

Q. 247 Of the following, which soil texture is highly susceptible to crusting?

(*a*) Clay loam (*b*) Silty clay

(*c*) Silty clay loam (*d*) Loam

Q. 248 Which mineral imparts black colour to the soil?

(*a*) Glauconite (*b*) Manganese oxide

(*c*) Lepidocrosite (*d*) Hematite

Q. 249 Which mineral imparts green colour to the soil?

(*a*) Glauconite (*b*) Manganese oxide

(*c*) Lepidocrosite (*d*) Hematite

Q. 250 What is the viscosity of water at 25°C?

(*a*) 0.0089 poise (*b*) 0.089 poise

(*c*) 0.89 poise (*d*) 8.9 poise

Q. 251 The soil moisture characteristic can be determined with the help of pressure plate apparatus in the broad range of suction _________:

(*a*) From 0 to 15 bar (*b*) From 1 to 15 bar

(*c*) Up to 1 bar (*d*) Up to –0.9 bar

Q. 252 Porus cup apparatus can be used for the determination of:

(*a*) Saturation water content

(*b*) Non-capillary water

(*c*) Drainable pore volume

(*d*) Both (*a*) and (*b*)

(*e*) All the above

Q. 253 Pressure plate can be used for the determination of:

(*a*) Field capacity

(*b*) Wilting point

(*c*) Non-capillary water

(*d*) Both (*a*) and (*b*)

(*e*) All the above

Q. 254 Of the following which soil property is more important in controlling water retention at low potentials (<-1 bar)?

(*a*) Structure (*b*) Texture

(*c*) Consistency (*d*) Colour

Q. 255 The volume of pore space filled with air when the soil is under a tension of __________ milli bar is called as air capacity or non-capillary porosity:

(*a*) 30 (*b*) 40

(*c*) 50 (*d*) 60

Q. 256 The diffusion of gases through dry soils of fine texture is higher than that through dry soils of coarse texture (True/False):

(*a*) True (*b*) False

Q. 257 The air-filled porosity in soil can be evaluated by:

(*a*) Air-space pycnometer (*b*) Air permeameter

(*c*) Gasometer (*d*) None

Q. 258 Air permeability of soil under field and laboratory conditions can be determined by

(*a*) Air permeameter (*b*) Gasometer

(*c*) Air-space pycnometer (*d*) Both (*a*) and (*b*)

(*e*) All the above

Q. 259 Unit of volumetric heat capacity is:

(*a*) J/m/sec/°C (*b*) m^2/sec

(*c*) J/m^3/°c (*d*) J/kg/°c

Q. 260 Black soils cover almost ____ % of the geographical area of India:

(*a*) 9 (*b*) 20

(*c*) 21 (*d*) 22

Q. 261 Red soils cover almost _____ % of the geographical area of India:

(*a*) 9 (*b*) 20

(*c*) 21 (*d*) 22

Q. 262 Plants generally start wilting at relative leaf water content of (RLWC) ______:

(*a*) 50–60 % (*b*) 60–75 %

(*c*) 75–85 % (*d*) 80–90 %

Q. 263 Soil texture influences:

(*a*) Cohesion (*b*) Adhesion

(*c*) Plasticity (*d*) All

Q. 264 Soil texture if assessed by feel method is done by:

(*a*) Ball formation (*b*) Ribbon formation

(*c*) Staining (*d*) All

Q. 265 Of soil feels very gritty by fingers and no ball is formed also no ribbon and stains are found then it is:

(*a*) Sand by texture (*b*) Loam by texture

(*c*) Clay by texture (*d*) All

Q. 266 Of a soil feels very gritty, forms very easily broken wall stechiness is little and no ribbon is formed then its texture is:

(*a*) Sand (*b*) Loamy sand

(*c*) Sandy loam (*d*) Loam

Q. 267 Of a soil feels moderately gritty, forms fairly firm ball which is easily broken, and it deficiency stains the finger with no ribbon formation it, texture is:

(*a*) Sand (*b*) Loamy sand

(*c*) Sandy loam (*d*) Loam

Q. 268 Of a soil feels neither very gritty nor very smooth and it forms firm ball, with stains is the fingers its texture is:

(*a*) Sand (*b*) Loamy sand

(*c*) Sandy loam (*d*) Loam

Q. 269 Of a soil feels smooth or stick buttery feel, forms firm ball stains in the fingers and has slight tendency to ribbon with flaky surface its texture is:

(*a*) Loam (*b*) Silt loam

(*c*) Clay loam (*d*) Clay

Q. 270 Of a soil feels slightly gritty, forms moderately hard ball when dry and stains in the finger and ribbons out on squeezing which also breaks easily it, texture is:

(*a*) Loam (*b*) Silt loam

(*c*) Clay loam (*d*) Clay

Q. 271 Of a soil feels very smooth, forms moderately hard ball when dry, stains in the finger and shows some fluking on ribbon surface similar to silt loam it, texture is:

(*a*) Silt loam (*b*) Silty clay loam

(*c*) Clay (*d*) Loam

Q. 272 Of a soil feels very smooth, forms hard ball which when dry cannot be crushed by fingers and stains definitely and when squared forms out into long ribbons, it, texture is:

(*a*) Clay loam (*b*) Loam

(*c*) Sand (*d*) Clay

Q. 273 Oxidation of organic carbon in soil sample can be obtained by treatment with:

(*a*) 10% H_2O_2 (*b*) 20% H_2O_2

(*c*) 30% H_2O_2 (*d*) 50% H_2O_2

Q. 274 H_2O_2 additions in soils result into:

(*a*) Frothing (*b*) Evolution of gas

(*c*) Change in soil Coles (*d*) All

Q. 275 Which of the following is a soil dispersing agent?

(*a*) Sodium fluoride

(*b*) Sodium chloride

(*c*) Sodium phosphate

(*d*) Sodium hexameta phosphate

Q. 276 The removal of free iron oxides requires use of which of the following reducing agent?

(*a*) Sodium dithionate (*b*) Sodium nitrate

(*c*) Both (*a*) and (*b*) (*d*) None

Q. 277 Present of $CaCO_3$ in soils makes dispersion:

(*a*) Easy (*b*) Difficult

(*c*) Same (*d*) No effect

Q. 278 The soil water suspension treated with sodium hexametaphosphate should be homogenized:

(*a*) For longer period

(*b*) For shorter period

(*c*) Not affected with time period

(*d*) None of the above

Q. 279 Use of water compared to non-polar liquids gives:

(*a*) Higher particle density value

(*b*) Lower particle density values

(*c*) Same particle density values

(*d*) None

Q. 280 Which of the following is/are non-polar liquid?

(*a*) Toluene (*b*) Xylene

(*c*) Carbon tetrachloride (*d*) All

Q. 281 Bulk density in soils can be estimated by:

(*a*) Core method (*b*) Excavation method

(*c*) Both (*d*) None

Q. 282 Soil core samples should not be taken from:

(*a*) Dry soils (*b*) Wet soils

(*c*) Both (*d*) None

Q. 283 Bulk density of a soil is influenced by:

(*a*) Green measuring (*b*) Soil conditioners

(*c*) Tillage (*d*) All

Q. 284 Which of the following is a simple notation of per cent total pore space?

(*a*) 100–% Solid space (*b*) 100–% Air space

(*c*) 100--% liquid space (*d*) All

Q. 285 Per cent solid space is soil is expressed by:

(*a*) Bulk density/Particle density × 100

(*b*) Particle density/Bulk density × 100

(*c*) (Particle density/Bulk density) × 100

(*d*) (Bulk density–Particle density) × 100

Q. 286 Per cent total pore spaces in soils are:

(*a*) Micro pores is soil

(*b*) Macro pores is soil

(*c*) Sum of micros macro pores is soil

(*d*) None

Q. 287 Who gave definition of air capacity of soil by field method?

(*a*) Kopecky (*b*) Block

(*c*) Single (*d*) None

Q. 288 Per cent capillary water can be estimated by:

(*a*) % Moisture in the soil at field capacity–% Moisture at hygroscopic co-efficient

(*b*) Mass of capillary water/Density of water

(*c*) % Capillary water × Moisture of oven dry soil/100

(*d*) Volume of capillary water × 100/Volume of soil

Q. 289 Per cent moisture of capillary water can be estimated by:

(*a*) % Moisture in the soil at field capacity–% Moisture at hygroscopic co-efficient

(*b*) Mass of capillary water/Density of water

(*c*) % Capillary water × Moisture of oven dry soil/100

(*d*) Volume of capillary water × 100/Volume of soil

Q. 290 Per cent volume of capillary water can be estimated by:

(*a*) % Moisture in the soil at field capacity–% Moisture at hygroscopic co-efficient

(*b*) Mass of capillary water/Density of water

(*c*) % Capillary water × Moisture of oven dry soil/100

(*d*) Volume of capillary water × 100/Volume of soil

Q. 291 Per cent capillary pore space can be estimated by:

(*a*) % Moisture in the soil at field capacity–% Moisture at hygroscopic co-efficient

(*b*) Mass of capillary water/Density of water

(*c*) % Capillary water × Moisture of oven dry soil/100

(*d*) Volume of capillary water × 100/Volume of soil

Q. 292 Air capacity of soil should be greater than _______ for all the crops except paddy:

(*a*) 10% (*b*) 20%

(*c*) 30% (*d*) 40%

Q. 293 The requirement of optimum air capacity of the soil by crops vary with:

(*a*) Types of plants

(*b*) Stages of plant development

(*c*) Types of soil

(*d*) All

Q. 294 The soil strength of a soil is positively related to its:

(*a*) Bulk density (*b*) Moisture content

(*c*) Both (*a*) and (*b*) (*d*) None

Q. 295 The soil strength is inversely related to:

(*a*) Bulk density (*b*) Moisture content

(*c*) Both (*a*) and (*b*) (*d*) None

Q. 296 The bulk density is a direct measure of:

(*a*) Soil strength (*b*) Soil compaction

(*c*) Moisture content (*d*) Soil solids

Q. 297 The bulk density of entire soil profile can be mapped by:

(*a*) Co–65 radiation (*b*) UV radiation

(*c*) α–rays (*d*) β–rays

Q. 298 Mineral particles in close contact with seeds other are held together by:

(*a*) Cohesion (*b*) Ligands

(*c*) Iron oxides (*d*) $CaCO_3$

(*e*) All the above

Q. 299 Aggregate size distribution affects:

(*a*) Air capacity (*b*) Water reduction

(*c*) Water movement (*d*) All

Q. 300 Aggregate size of soils are affected by:

(*a*) Tillage (*b*) Cropping

(*c*) Cultural practices (*d*) All

Q. 301 Soils having higher proportion of large size aggregate are:

(*a*) Less productive

(*b*) More productive

(*c*) Neither more nor less productive

(*d*) None

Q. 302 Soils with higher water stable aggregate will have a high proportion of:

(*a*) Small pores (*b*) Large pores

(*c*) Very small pore (*d*) All

Q. 303 High infiltration in soils is influenced by:

(*a*) Small pores (*b*) Large pores

(*c*) Very small pore (*d*) All

Q. 304 Effect of tillage, cultural and manorial practices on soil aggregate can be evaluated by determining:

(*a*) MWD (*b*) GMD

(*c*) log GMD (*d*) % Aggregate

(*e*) All the above

Q. 305 The hydraulic conductivity of soil is affected by:

(*a*) Temperature (*b*) Pressure

(*c*) Both (*d*) None

Q. 306 Intrinsic permeability to water is related to:

(*a*) Hydraulic conductivity (*b*) Infiltration

(*c*) Bulk density (*d*) Particle density

Q. 307 When a soil is having hydraulic conductivity > 25.0 cm hr^{-1}, it, permeability will be:

(*a*) Slow (*b*) Moderate

(*c*) Rapid (*d*) Very rapid

Q. 308 Field capacity soil moisture constant signifies the:

(*a*) Upper level of plant available water

(*b*) Lower level of plant available water

(*c*) Both (*a*) and (*b*)

(*d*) None of the above

Q. 309 Which of the following is a close approximation of field capacity?

(*a*) Moisture equivalent

(*b*) Capillary conductivity

(*c*) Capillary volume

(*d*) Hygroscopic coefficient

Q. 310 Which of the following notation is used to relate field capacity with moisture equivalent?

(*a*) FC = 2.62 + MC × 0.865 (*b*) WC = ME/1.84

(*c*) HC = WC × 0.684 (*d*) None

Q. 311 Which of the following notation is used to relate water content with moisture equivalent?

(*a*) FC = 2.62 + MC × 0.865 (*b*) WC = ME/1.84

(*c*) HC = WC × 0.684 (*d*) None

Q. 312 Which of the following notation is used to relate hygroscopic coefficient with water content?

(*a*) FC = 2.62 + MC × 0.865 (*b*) WC = ME/1.84

(*c*) HC = WC × 0.684 (*d*) None

Q. 313 The concept of incipient wilting was first described by:

(*a*) Furr and Reeve

(*b*) Veihmeyers Hendrickson

(*c*) Milk

(*d*) Bouyoucos

Q. 314 A plant having larges top/root ratio will be:

(*a*) Less affected by moisture stress

(*b*) More affected by moistures stress

(*c*) Not affected

(*d*) None of the above

Q. 315 There permanent wilting percentage vary with:

(*a*) Soil texture (*b*) Soil structure

(*c*) Mineralogical composition (*d*) All

Q. 316 Plant available water is the difference between:

(*a*) Field capacity and Permanent wilting point

(*b*) Permanent wilting point and Field capacity

(*c*) Hydraulic conductivity and Permanent wilting point

(*d*) Permanent wilting point and Hydraulic conductivity

Q. 317 Soil moisture tension increase with the:

(*a*) Increase in soil moisture content

(*b*) Decrease in soil

(*c*) Both (*a*) and (*b*)

(*d*) None of the above

Q. 318 If tensiometer is installed in the evening, when one can take measurement on soil moisture tension?

(*a*) Next morning (*b*) Next evening

(*c*) Same night (*d*) Over night

Q. 319 Relationship between soil moisture tension and soil moisture content depends upon

(*a*) Soil texture (*b*) Organic portion

(*c*) Mineralogical composition (*d*) All

Q. 320 Tensiometer measurement is a measurement of:

(*a*) Matric potential

(*b*) Energy status of water

(*c*) Soil moisture potential

(*d*) All

Q. 321 The porus cup in a mercury manometer is made up of:

(*a*) Sand (*b*) Kaolin suspension

(*c*) Soil: water suspension (*d*) Mercury

Q. 322 Tensiometer measurements are not suitable for:

(*a*) Sandy soils (*b*) Loamy soils

(*c*) Silty soils (*d*) Clayey soils

Q. 323 Scheduling of irrigation is crops should be based on:

(*a*) Soil moisture content (*b*) Soil moisture regime

(*c*) Both (*a*) and (*b*) (*d*) None

Q. 324 Which indices of soil structure will be highest for FYM?

(*a*) Water Soluble Aggregate (%)

(*b*) Mean Weight Diameter

(*c*) Geometric Mean Diameter

(*d*) Cumulative Mean Weight Diameter

Q. 325 Which indices of soil structure will be highest for manure?

(*a*) Water Soluble Aggregate (%)

(*b*) Mean Weight Diameter

(*c*) Geometric Mean Diameter

(*d*) Cumulative Mean Weight Diameter

Q. 326 Which indices of soil structure will be highest for SSP?

(*a*) Water Soluble Aggregate (%)

(*b*) Mean Weight Diameter

(*c*) Geometric Mean Diameter

(*d*) Cumulative Mean Weight Diameter

Q. 327 1 bar is equal to:

(*a*) 10^4 dynes cm^{-2}
(*b*) 10^5 dynes cm^{-2}
(*c*) 10^6 dynes cm^{-2}
(*d*) None

Q. 328 1 bar is equal to:

(*a*) 1 KPU
(*b*) 10 KPU
(*c*) 100 KPU
(*d*) None

Q. 329 Which gas has more percent volume composition in swampy rice land?

(*a*) N_2
(*b*) O_2
(*c*) CO_2
(*d*) H_2

Q. 330 Coefficient of linear extensibility is also denoted by:

(*a*) PVC
(*b*) Swell index
(*c*) Both (*a*) and (*b*)
(*d*) None

Q. 331 Soil crusting is related to:

(*a*) Soil composition
(*b*) Bulk density
(*c*) Soil strength
(*d*) All the above
(*e*) None of the above

Q. 332 Soil crusts are formed due to:

(*a*) High Exchangeable Sodium Percentage
(*b*) Poor structure
(*c*) Low organic matter
(*d*) All the above

Q. 333 Content of CO_2 in soils is:.

(*a*) Higher
(*b*) Lower
(*c*) Same
(*d*) No change than atmospheric CO_2

Q. 334 Size of which of the primary particles of soil is same in both ISSS and USDA system:

(*a*) Sand
(*b*) Silt
(*c*) Clay
(*d*) Both (*a*) and (*b*)

Q. 335 Under USDA system of soil textural classification, size of silt is:

(*a*) <0.05 mm (*b*) >0.05 mm
(*c*) 0.05 to 0.002 mm (*d*) < 0.0002 mm

Q. 336 Size of clay under ISSS system is:

(*a*) < 0.02 mm (*b*) < 0.2 mm
(*c*) < 0.002 mm (*d*) < 0.0002 mm

Q. 337 In gravimetric method of soil moisture estimation, moist soil is oven dried at:

(*a*) 10 °C (*b*) 105 °C
(*c*) 140 °C (*d*) 120 °C

Q. 338 Which of the soil materials listed below is likely to have the smallest bulk density?

(*a*) Sand (*b*) Silt loam
(*c*) Peat (*d*) Sandy loam

Q. 339 If large amounts of sand are incorporated into a flower bed the bulk density would:

(*a*) Increase (*b*) Decrease
(*c*) Not change (*d*) None

Q. 340 Compaction of a soil due to golf cart traffic would cause the bulk density to:

(*a*) Increase (*b*) Decrease
(*c*) Not change (*d*) Be unity

Q. 341 Compaction of a soil causes the water holding capacity to:

(*a*) Increase (*b*) Decrease
(*c*) Not change (*d*) Be unity

Q. 342 The total pore space in a soil is equal to the percent solids:

(*a*) The particle density of the sand
(*b*) The percent water by volume when a soil is saturated
(*c*) The bulk density of the sand
(*d*) The particle density of the soil

Q. 343 An increase in soil water content causes soil strength to:

(*a*) Increase (*b*) Decrease

(*c*) Not change (*d*) None

Q. 344 The cohesive and adhesive properties of water are due to the:

(*a*) Slight polarity of the individual molecules

(*b*) Capillary forces

(*c*) Gravity

(*d*) Surface tension

Q. 345 As soil moisture increases the soil air:

(*a*) Increases (*b*) Decreases

(*c*) Stays the same (*d*) None

Q. 346 Water has a high specific heat, which means:

(*a*) A relatively small amount of heat is required to change its temperature

(*b*) A relatively large amount of heat is required to change its temperature

(*c*) No specific amount of heat is required to change temperature

(*d*) Both (*a*) and (*b*)

(*e*) All the above

Q. 347 The temperature of a wet soil increases more slowly than a dry soil in the spring because of:

(*a*) The high specific heat of water

(*b*) The high conductivity of water

(*c*) Wetness does not influence warming of soil in the spring

(*d*) Both (*a*) and (*b*)

(*e*) All the above

Q. 348 Which soil separate has the greatest total water holding capacity?

(*a*) Sand (*b*) Clay

(*c*) Silt (*d*) None

Q. 349 Gravitational water in soil move through the:

(*a*) Micropores (*b*) Macropores

(*d*) Both (*a*) and (*b*) (*d*) None

Q. 350 Water held against the force of gravity in micropores is:

(*a*) Heavy water (*b*) Gravitational water

(*c*) Capillary water (*d*) Hygroscopic water

Q. 351 Water tightly held in thin films around soil particles is:

(*a*) Hygroscopic water (*b*) Gravitational water

(*c*) Capillary water (*d*) Heavy water

Q. 352 Capillary rise of water is due to the forces of:

(*a*) Cohesion (*b*) Adhesion

(*c*) Cohesion and adhesion (*d*) Matric potential

Q. 353 Which type of water includes that which is available to plants?

(*a*) Hygroscopic water (*b*) Gravitational water

(*c*) Capillary water (*d*) All

Q. 354 Capillary water in the soil is that which is found in the:

(*a*) Macropores (*b*) Micropores

(*c*) Root channels (*d*) Both (*a*) and (*b*)

Q. 355 Water holding capacity of a soil is affected by:

(*a*) Texture (*b*) Organic matter

(*c*) Structure (*d*) All of the above

Q. 356 The osmotic potential of soil water is affected by:

(*a*) Salt content (*b*) Barometric pressure

(*c*) Clay content (*d*) All

Q. 357 When soil is saturated the dominant force affecting water movement is:

(*a*) Chemical bonding (*b*) Capillary tension

(*c*) Gravity (*d*) Van der Waal's force

Q. 358 The water held between –1/3 and –15 bar is called:

(*a*) Gravitational water (*b*) Plant available water

(*c*) Wilting point (*d*) Field capacity

Q. 359 Transpiration ratio indicates:

(*a*) The amount of water transpired by a crop divided by 10

(*b*) The amount of water required to produce 1 pound of dry weight of plant material

(*c*) The amount of water in a bushel of corn

(*d*) Both (*a*) and (*c*)

(*e*) All the above

Q. 360 A layer of sand in a silt loam soil:

(*a*) Restricts saturated flow

(*b*) Has no effect on water movement

(*c*) Restricts unsaturated flow downward

(*d*) Both (*a*) and (*c*)

Q. 361 In order for water to move from the silt loam through the sand layer the moisture content of the silt loam must be

(*a*) Air dry (*b*) Near saturation

(*c*) At the wilting point (*d*) Both (*a*) and (*c*)

(*e*) All the above

Q. 362 A clay layer in the silt loam would restrict water movement because

(*a*) Capillary forces are greater in the silt loam

(*b*) Small pores in the clay transmit water slowly

(*c*) Larger pores in the clay restrict movement

(*d*) Both (*a*) and (*b*)

(*e*) All the above

Q. 363 At field capacity, a root channel would be filled with:

(*a*) Water (*b*) Air

(*c*) Soil (*d*) Both (*a*) and (*b*)

Q. 364 Gravitational water is not considered available to plants because:

(*a*) It moves out of the root zone quickly in well drained soils

(*b*) No oxygen is available for root respiration while the soil is saturated

(*c*) Both of the above

(*d*) None of the above

Q. 365 Horizontal and upward movement of water in soil is nearly as great as downward movement when the soil is:

(*a*) Saturated (*b*) Unsaturated

(*c*) Partly saturated (*d*) Submerged

Q. 366 More hygroscopic water would be present in:

(*a*) Sand (*b*) Silt

(*c*) Clay (*d*) Both (*b*) and (*c*)

Q. 367 One indicator of wetland soils is:

(*a*) Silt loam textural class

(*b*) Gray colors within one foot of the surface

(*c*) Red colors in the B horizon

(*d*) Both (*a*) and (*c*)

(*e*) All the above

Q. 368 The criteria required to define for jurisdictional wetlands are hydrology, hydric soils and:

(*a*) Water standing on the surface

(*b*) Clay subsoils

(*c*) Wetland vegetation

(*d*) All of the above

Q. 369 Wetland soils usually have:

(*a*) Greater amounts of organic matter than well drained soils

(*b*) Less organic matter than well drained soils

(*c*) More oxygen than well drained soils

(*d*) Undisturbed structures

Q. 370 When saturated conditions occur in a soil above an impermeable soil layer such as clay, it is referred to as:

(*a*) A perched water table (*b*) The capillary fringe

(*c*) The aquic zone (*d*) The udic zone

Q. 371 The plant nutrients which are of concern when they are present in excess amounts in lakes, rivers and estuaries are:

(*a*) Nitrogen (*b*) Phosphorus

(*c*) Sulphur (*d*) Both N and P

Q. 372 Nonpoint source pollution may come from:

(*a*) Sewage outfalls

(*b*) Runoff from farms and urban areas

(*c*) Drainage from farms and urban areas

(*d*) Both (*b*) and (*c*)

(*e*) All the above

Q. 373 A wet zone in soils above the water table resulting from capillary rise is called the:

(*a*) Capillary fringe (*b*) Perched water table

(*c*) Aquafina (*d*) Both (*a*) and (*b*)

Q. 374 If a sandy loam soil and a clay soil are both at field capacity, the percent moisture will be:

(*a*) Greater in the sandy loam (*b*) Greater in the clay

(*c*) The same in both soils (*d*) None

Q. 375 Equal amounts of water applied to a sandy loam soil and a clay loam soil will percolate to a greater depth in:

(*a*) The sandy loam

(*b*) The clay loam

(*c*) There would be no difference

(*d*) None

Q. 376 The inner film of water around a soil particle is held by:

(*a*) Cohesion (*b*) Adhesion

(*c*) Chemical bond (*d*) Both (*a*) and (*b*)

Q. 377 Hygroscopic water can be removed from soil by:

(*a*) Drainage tile (*b*) Air drying

(*c*) Oven drying (*d*) Both (*b*) and (*c*)

Q. 378 Which of the following is not a kind of soil structure?

(*a*) Crumb (*b*) Platy

(*c*) Angular blocky (*d*) Flat crumb

Q. 379 Identify particle sizes by least to greatest or fine to coarse:

(*a*) Sand, Silt, Clay

(*b*) Clay, Sand, Silt

(*c*) Silt, Sand, Clay

(*d*) Clay, Silt, Sand

Q. 380 A soil textural triangle measures:

(*a*) Percentage of sand, silt, clay

(*b*) Mineral levels in a soil sample

(*c*) Percent slope of a land class

(*d*) Percentage of soil erosion on a given location

Q. 381 Which soil particles have the ability to have excessive drainage?

(*a*) Silt (*b*) Clay

(*c*) Sand (*d*) None of the above

Q. 382 In level or low spots, black topsoil with gray coloured subsoil, the drainage class would be ________:

(*a*) Somewhat poorly drained

(*b*) Moderately well drained

(*c*) Well drained

(*d*) Very poorly drained

Q. 383 The average composition of soil contains:

(*a*) 53% parent material, 7% organic matter, 5% air, 35% water

(*b*) 25% air, 25% water, 5% mineral matter, 45% organic matter

(*c*) 25% organic matter, 25% water, 5% air, 45% mineral matter

(*d*) 25% air, 25% water, 5% organic matter, 45% mineral matter

Q. 384 Relative lightness or darkness in Munsell colour chart indicates:

(*a*) Hue (*b*) Value

(*c*) Chroma (*d*) All

Q. 385 Osmotic potential is also called as:

(*a*) Solute potential

(*b*) Matric potential

(*c*) Pressure potential

(*d*) Gravitational potential

Q. 386 Gravitational potential is independent of chemical potential but dependent on:

(*a*) Vertical location (*b*) Density of water

(*c*) Both (*a*) and (*b*) (*d*) None

Q. 387 Solute and matric potential are collectively called as:

(*a*) Osmotic potential (*b*) Matric potential

(*c*) Gravitational potential (*d*) All

Q. 388 Soil water potential always has:

(*a*) Positive sign (*b*) Negative sign

(*c*) No sign (*d*) None

Q. 389 Soil water suction always has

(*a*) Positive sign (*b*) Negative sign

(*c*) No sign (*d*) None

ANSWERS

1	(*a*)	23	(*a*)	45	(*d*)	67	(*a*)
2	(*a*)	24	(*b*)	46	(*d*)	68	(*c*)
3	(*b*)	25	(*c*)	47	(*b*)	69	(*b*)
4	(*b*)	26	(*a*)	48	(*c*)	70	(*a*)
5	(*c*)	27	(*d*)	49	(*d*)	71	(*b*)
6	(*d*)	28	(*b*)	50	(*a*)	72	(*a*)
7	(*b*)	29	(*c*)	51	(*c*)	73	(*d*)
8	(*b*)	30	(*a*)	52	(*b*)	74	(*c*)
9	(*c*)	31	(*b*)	53	(*a*)	75	(*b*)
10	(*d*)	32	(*a*)	54	(*b*)	76	(*d*)
11	(*a*)	33	(*c*)	55	(*a*)	77	(*a*)
12	(*c*)	34	(*d*)	56	(*c*)	78	(*c*)
13	(*a*)	35	(*b*)	57	(*a*)	79	(*b*)
14	(*b*)	36	(*b*)	58	(*c*)	80	(*a*)
15	(*d*)	37	(*e*)	59	(*b*)	81	(*b*)
16	(*a*)	38	(*a*)	60	(*a*)	82	(*a*)
17	(*c*)	39	(*b*)	61	(*c*)	83	(*e*)
18	(*b*)	40	(*d*)	62	(*d*)	84	(*c*)
19	(*a*)	41	(*c*)	63	(*d*)	85	(*b*)
20	(*d*)	42	(*b*)	64	(*a*)	86	(*a*)
21	(*a*)	43	(*a*)	65	(*b*)	87	(*d*)
22	(*b*)	44	(*d*)	66	(*a*)	88	(*c*)

89	(*b*)	119	(*b*)	149	(*c*)	179	(*c*)
90	(*c*)	120	(*a*)	150	(*d*)	180	(*a*)
91	(*b*)	121	(*c*)	151	(*d*)	181	(*b*)
92	(*c*)	122	(*a*)	152	(*c*)	182	(*a*)
93	(*b*)	123	(*a*)	153	(*b*)	183	(*c*)
94	(*c*)	124	(*b*)	154	(*c*)	184	(*b*)
95	(*a*)	125	(*d*)	155	(*d*)	185	(*c*)
96	(*c*)	126	(*c*)	156	(*a*)	186	(*b*)
97	(*b*)	127	(*b*)	157	(*b*)	187	(*b*)
98	(*d*)	128	(*a*)	158	(*c*)	188	(*a*)
99	(*a*)	129	(*b*)	159	(*d*)	189	(*a*)
100	(*b*)	130	(*a*)	160	(*d*)	190	(*c*)
101	(*a*)	131	(*c*)	161	(*c*)	191	(*c*)
102	(*d*)	132	(*c*)	162	(*a*)	192	(*d*)
103	(*c*)	133	(*b*)	163	(*d*)	193	(*b*)
104	(*a*)	134	(*a*)	164	(*c*)	194	(*a*)
105	(*b*)	135	(*b*)	165	(*b*)	195	(*c*)
106	(*c*)	136	(*a*)	166	(*b*)	196	(*c*)
107	(*a*)	137	(*c*)	167	(*a*)	197	(*a*)
108	(*c*)	138	(*d*)	168	(*a*)	198	(*c*)
109	(*a*)	139	(*c*)	169	(*c*)	199	(*a*)
110	(*b*)	140	(*b*)	170	(*a*)	200	(*a*)
111	(*c*)	141	(*a*)	171	(*b*)	201	(*a*)
112	(*d*)	142	(*b*)	172	(*c*)	202	(*c*)
113	(*d*)	143	(*b*)	173	(*a*)	203	(*b*)
114	(*d*)	144	(*d*)	174	(*a*)	204	(*c*)
115	(*a*)	145	(*a*)	175	(*d*)	205	(*b*)
116	(*b*)	146	(*c*)	176	(*a*)	206	(*b*)
117	(*a*)	147	(*b*)	177	(*a*)	207	(*a*)
118	(*c*)	148	(*a*)	178	(*a*)	208	(*a*)

209 (*b*)
210 (*d*)
211 (*c*)
212 (*c*)
213 (*a*)
214 (*c*)
215 (*a*)
216 (*a*)
217 (*d*)
218 (*b*)
219 (*a*)
220 (*a*)
221 (*c*)
222 (*c*)
223 (*b*)
224 (*d*)
225 (*b*)
226 (*a*)
227 (*e*)
228 (*c*)
229 (*a*)
230 (*a*)
231 (*d*)
232 (*b*)
233 (*c*)
234 (*d*)
235 (*d*)
236 (*b*)
237 (*c*)
238 (*d*)
239 (*b*)
240 (*b*)
241 (*c*)
242 (*a*)
243 (*c*)
244 (*d*)
245 (*c*)
246 (*d*)
247 (*c*)
248 (*b*)
249 (*a*)
250 (*a*)
251 (*a*)
252 (*e*)
253 (*d*)
254 (*b*)
255 (*c*)
256 (*a*)
257 (*a*)
258 (*d*)
259 (*c*)
260 (*d*)
261 (*b*)
262 (*c*)
263 (*d*)
264 (*d*)
265 (*a*)
266 (*b*)
267 (*c*)
268 (*d*)
269 (*b*)
270 (*c*)
271 (*b*)
272 (*d*)
273 (*c*)
274 (*d*)
275 (*d*)
276 (*c*)
277 (*b*)
278 (*b*)
279 (*a*)
280 (*d*)
281 (*c*)
282 (*c*)
283 (*d*)
284 (*a*)
285 (*a*)
286 (*c*)
287 (*a*)
288 (*a*)
289 (*c*)
290 (*b*)
291 (*d*)
292 (*a*)
293 (*d*)
294 (*a*)
295 (*b*)
296 (*b*)
297 (*a*)
298 (*e*)
299 (*d*)
300 (*d*)
301 (*b*)
302 (*b*)
303 (*b*)
304 (*e*)
305 (*a*)
306 (*a*)
307 (*d*)
308 (*a*)
309 (*a*)
310 (*a*)
311 (*b*)
312 (*c*)
313 (*a*)
314 (*b*)
315 (*d*)
316 (*a*)
317 (*b*)
318 (*a*)
319 (*d*)
320 (*d*)
321 (*b*)
322 (*d*)
323 (*b*)
324 (*a*)
325 (*a*)
326 (*a*)
327 (*c*)
328 (*c*)

329	(*a*)	344	(*a*)	359	(*b*)	374	(*b*)
330	(*c*)	345	(*b*)	360	(*c*)	375	(*a*)
331	(*d*)	346	(*b*)	361	(*b*)	376	(*b*)
332	(*d*)	347	(*a*)	362	(*b*)	377	(*c*)
333	(*a*)	348	(*b*)	363	(*b*)	378	(*d*)
334	(*d*)	349	(*b*)	364	(*c*)	379	(*d*)
335	(*c*)	350	(*c*)	365	(*b*)	380	(*a*)
336	(*c*)	351	(*a*)	366	(*c*)	381	(*c*)
337	(*b*)	352	(*c*)	367	(*b*)	382	(*d*)
338	(*c*)	353	(*c*)	368	(*c*)	383	(*d*)
339	(*a*)	354	(*b*)	369	(*a*)	384	(*b*)
340	(*a*)	355	(*d*)	370	(*a*)	385	(*a*)
341	(*b*)	356	(*a*)	371	(*d*)	386	(*c*)
342	(*c*)	357	(*c*)	372	(*b*)	387	(*a*)
343	(*b*)	358	(*b*)	373	(*a*)	388	(*b*)
						389	(*a*)

Chapter 10

Soil and Water Conservation

Q. 1 Erosion created by the activities of man and sometimes by animals is called as:

(*a*) Natural geological erosion (*b*) Geologic erosion

(*c*) Accelerated soil erosion (*d*) All

Q. 2 Scattering of detached soil particles by the impact of rain drops is called:

(*a*) Splash erosion (*b*) Sheet erosion

(*c*) Gully erosion (*d*) Ravines

Q. 3 Removal by rain of a very thin layer of soil from the entire surface of large area is called:

(*a*) Splash erosion (*b*) Sheet erosion

(*c*) Rill erosion (*d*) Ravines

Q. 4 Sheet erosion is observed on:

(*a*) Deeply sloping land

(*b*) Moderately sloping land

(*c*) Slopy land

(*d*) Gently sloping land

Q. 5 Muddy run-off from the field is an indication of:

(*a*) Splash erosion (*b*) Sheet erosion

(*c*) Rill erosion (*d*) Ravines

Q. 6 Formation of small channels a few centimeters deep al over the field along the watercourse is observed in:

(*a*) Rill erosion (*b*) Slight gully erosion

(*c*) Moderate gully erosion (*d*) Ravines

Q. 7 Run-off and soil loss can be calculated by:

(*a*) Ramser's formula

(*b*) Chepil equation

(*c*) Chepil and Woodruff equation

(*d*) All

Q. 8 Device used to study the run-off and soil loss under laboratory conditions:

(*a*) Run-off plots (*b*) Multislot device

(*c*) Standing wavefumes (*d*) Rainfall simulator

Q. 9 The quantity of soil materials of a given size in the run-off varies as the ______ power of the speed of run-off:

(*a*) Third (*b*) Fourth

(*c*) Fifth (*d*) Sixth

Q. 10 Erosion permitting crops is/are:

(*a*) Maize (*b*) Wheat

(*c*) Bajra (*d*) All

Q. 11 Erosion resisting crops is/are:

(*a*) Grasses (*b*) Legumes

(*c*) Cereals (*d*) Both (*a*) and (*b*)

(*d*) All the above

Q. 12 Soils more susceptible to water erosion are:

(*a*) Coarse textured soils

(*b*) Medium textured soils

(*c*) Fine textured soils

(*d*) All

Q. 13 The ratio of dispersion ratio to colloidal moisture equivalent ratio is:

(*a*) Erosion index (*b*) Erosion ratio

(*c*) Soil detachability (*d*) Soil erosivity

Q. 14 In Universal soil loss equation, A=RKLSCP, K denotes:

(*a*) Soil conservation practices (*b*) Soil roughness
(*c*) Soil erodability (*d*) Rainfall factor

Q. 15 In Universal soil loss equation, A=RKLSCP, P denotes:

(*a*) Soil conservation practices (*b*) Soil roughness
(*c*) Soil erodability (*d*) Rainfall factor

Q. 16 Growing of erosion resisting and erosion permitting crops on alternate strips of suitable width along the contour and across the slope is called as:

(*a*) Contour cropping (*b*) Contour farming
(*c*) Contour strip cropping (*d*) Strip cropping

Q. 17 Contour bunding is also called as:

(*a*) Level terraces (*b*) Ridge-type terraces
(*c*) Absorption type terraces (*d*) All

Q. 18 Strong dusty winds are locally called as:

(*a*) Aeolian (*b*) Loo
(*c*) Loess (*d*) None

Q. 19 Minimum wind speed required to initiate the movement of most erodable soil particles is:

(*a*) 10 km (*b*) 12 km
(*c*) 14 km (*d*) 16 km

Q. 20 Diameter of particles that is most susceptible to wind erosion is:

(*a*) 0.1 mm (*b*) 0.2 mm
(*c*) 0.02 mm (*d*) 0.01 mm

Q. 21 Dominant type of wind erosion in soil is:

(*a*) Saltation (*b*) Surface creep
(*c*) Suspension (*d*) All

Q. 22 Size range of soil particles in saltation type of wind erosion is:

(*a*) Less than 0.05 mm (*b*) 0.5–1.0 mm

(*c*) 0.5–2.0 mm (*d*) 0.05–0.5 mm

Q. 23 Size range of soil particles in suspension type of wind erosion is:

(*a*) Less than 0.05 mm (*b*) 0.5–1.0 mm

(*c*) 0.5–2.0 mm (*d*) 0.05–0.5 mm

Q. 24 Size range of soil particles in surface creep type of wind erosion is:

(*a*) Less than 0.05 mm (*b*) 0.5–1.0 mm

(*c*) 0.5–2.0 mm (*d*) 0.05–0.5 mm

Q. 25 In surface creep wind erosion, soil particles are moved primarily by the action of

(*a*) Wind (*b*) Saltation

(*c*) Suspension (*d*) All

Q. 26 Most important factors that influence the erodability of soil by wind:

(*a*) Soil texture (*b*) Soil structure

(*c*) Soil water status (*d*) All

Q. 27 Wind erosion is calculated by:

(*a*) Ramser's formula

(*b*) Chepil equation

(*c*) Chepil and Woodruff equation

(*d*) All

Q. 28 In Chepil and Woodruff's equation of soil loss calculation by wind erosion, E = f (ICKLV), I denotes:

(*a*) Soil erodability (*b*) Rainfall intensity

(*c*) Soil roughness (*d*) Soil type

Q. 29 In Chepil and Woodruff's equation of soil loss calculation by wind erosion, E = f (ICKLV), K denotes:

(*a*) Soil erodability (*b*) Rainfall intensity

(*c*) Soil roughness (*d*) Soil type

Q. 30 The process that lowers the current and/or potential capability of soil to produce goods or services is called:

(*a*) Soil degradation (*b*) Land degradation

(*c*) Mass degradation (*d*) All

Q. 31 The desertification was first used by:

(*a*) Chepil and Woodruff (*b*) Weischner

(*c*) Abreville (*d*) None

Q. 32 The process that transforms the productive land into unproductive land:

(*a*) Desertification (*b*) Desertization

(*c*) Mass movement (*d*) Both (*a*) and (*b*)

(*e*) All the above

Q. 33 Extension of tropical desert landscapes and landforms to area where they did not occur in the recent past is called as:

(*a*) Desertification (*b*) Desertization

(*c*) Mass movement (*d*) Both (*a*) and (*b*)

(*e*) All the above

Q. 34 In north-eastern hill regions of India, shifting cultivation is also called as

(*a*) Jhum cultivation (*b*) Jhuming cultivation

(*c*) Both (*a*) and (*b*) (*d*) None

Q. 35 Total number of land capability classes in the land capability classification scheme:

(*a*) Six (*b*) Seven

(*c*) Eight (*d*) Nine

Q. 36 Land capability classes are subdivided in to land capability subclasses based on their:

(*a*) Major characteristics

(*b*) Kind of managements

(*c*) Kind of limitations

(*d*) Both (*a*) and (*b*)

(*e*) All the above

Q. 37 Total number of classes in the group, land suitable for cultivation is:

(*a*) Eight (*b*) Four

(*c*) Six (*d*) Two

Q. 38 In land capability classification scheme, the classes V, VI and VII are suitable for:

(*a*) Cultivation

(*b*) Pasture and grazing

(*c*) Wildlife and water shed management

(*d*) All

Q. 39 Land capability subclasses are further subdivided into land capability units on the basis of:

(*a*) Major characteristics

(*b*) Kind of managements

(*c*) Kind of limitations

(*d*) Both (*a*) and (*b*)

(*e*) All the above

Q. 40 Rills with more than ______ cm depth are generally called as gullies:

(*a*) 15 (*b*) 20

(*c*) 25 (*d*) 30

Q. 41 Of the following, which one is important with regards to soil erosion by water?

(*a*) Intensity of rainfall (*b*) Duration of rain fall

(*c*) Amount of rainfall (*d*) Frequency of rainfall

Q. 42 If the velocity of water is doubled, its erosion power increases ___ times, and carrying capacity increases___ times:

(*a*) 2 ... 16 (*b*) 4 ... 16

(*c*) 4 ... 32 (*d*) 4 ... 64

Q. 43 The most serious sheet erosion occurs in:

(*a*) Red and Black soils

(*b*) Red and Alluvial soils

(*c*) Black and Alluvial soils

(*d*) Red and laterite soils

Q. 44 Wind erosion is mainly due to:

(*a*) A strong wind regime

(*b*) Low atmospheric humidity

(*c*) High solar radiation

(*d*) Single grain structure of sandy soils

(*e*) All the above

Q. 45 Which of the following mechanical soil conservation measures is more suitable for fruit trees or other plantation crops on steep slopes?

(*a*) Contour bunds (*b*) Half moon terraces

(*c*) Graded bunds (*d*) Bench terraces

Q. 46 Land degradation first starts with the:

(*a*) Reduction in vegetative cover

(*b*) Exposing the soil surface to accelerated erosion

(*c*) Reduction in soil organic matter and nutrient content

(*d*) All the above

Q. 47 Tillage is a process to optimize conditions for germination bringing seedling establishment and crop growth and is

(*a*) A physical

(*b*) Chemical

(*c*) Biological soil manipulation

(*d*) All the above

Q. 48 The primary objectives of tillage is:

(*a*) Seedbed preparation

(*b*) Provision of a good medium for plant roots

(*c*) Water infiltration and retention

(*d*) Erosion and weed control

Q. 49 Conservation tillage can be defined as:

(*a*) A crop planting system that allows minimum disturbance of the soil to allow seeds to be sown while ensuring maintenance of crop residues on the surface

(*b*) A tillage operation is mainly done to prevent the weed population

(*c*) A tillage operation is mainly done to prepare a fine seed bed

(*d*) All the above

Q. 50 What is important reason of soil erosion?

(*a*) Erosion removes topsoil

(*b*) Reduces levels of soil organic matter

(*c*) It contributes to the breakdown of soil structure

(*d*) All the above

Q. 51 What are signs of wind erosion?

(*a*) Dust clouds

(*b*) Soil accumulation along fencelines or snowbanks

(*c*) A drifted appearance of the soil surface

(*d*) All the above

Q. 52 What are signs of water erosion?

(*a*) Small rills and channels on the soil surface

(*b*) Soil deposited at the base of slopes

(*c*) Sediment in streams, lakes, and reservoirs

(*d*) All the above

Q. 53 What are signs of water erosion?

(*a*) Pedestals of soil supporting pebbles and plant material

(*b*) Convex landscape positions

(*c*) Both (*a*) and (*b*)

(*d*) None of the above

Q. 54 Long-term soil erosion results in

(*a*) Persistent and large gullies

(*b*) Exposure of lighter coloured subsoil at the surface

(*c*) Poorer plant growth

(*d*) All the above

Q. 55 How can soil erosion be measured?

(*a*) Visual

(*b*) Physical

(*c*) Chemical

(*d*) Biological indicators

(*e*) All the above

Q. 56 Which of the following is a visual indicators of soil erosion?

(*a*) Comparisons of aerial photographs taken over time

(*b*) Presence of moss and algae (crypotogams) crusts in desert or arid soils

(*c*) Changes in soil horizon thickness

(*d*) Deposition of soil at field boundaries

(*e*) All the above

Q. 57 Which of the following is a physical indicator of soil erosion?

(*a*) Measurements of aggregate stability

(*b*) Increasing depth of channels and gullies

(*c*) Both (*a*) and (*b*)

(*d*) None of the above

Q. 58 Which of the following is a chemical indicator of soil erosion?

(*a*) Decreases in soil organic matter content

(*b*) Increases in calcium carbonate content at the surface

(*c*) Provided greater content exists in subsurface layers

(*d*) Changes in cation-exchange capacity (CEC)

(*e*) All the above

Q. 59 Which of the following is a biological indicator of soil erosion?

(*a*) Decreased microbial biomass

(*b*) Lower rate of respiration

(*c*) Slower decomposition of plant residues

(*d*) All the above

Q. 60 What causes water erosion?

(*a*) Lack of protection against raindrop impact

(*b*) Decreased aggregate stability

(*c*) Long and steep slopes

(*d*) Intense rainfall or irrigation events when plant or

(*e*) Residue cover is at a minimum

(*f*) Decreased infiltration by compaction or other means

Q. 61 What causes mechanical erosion?

(*a*) Removal by harvest of root crops

(*b*) Tillage and cultivation practices that move soil

(*c*) Downslope

(*d*) All the above

Q. 62 What causes wind erosion?

(*a*) Exposed surface soil during critical periods of the year

(*b*) Occurrence of wind velocities that are sufficient to lift

(*c*) Individual soil particles

(*d*) Long, unsheltered, smooth soil surfaces

(*e*) All the above

Q. 63 Soil erosion can be avoided by:

(*a*) Maintaining a protective cover on the soil

(*b*) Creating a barrier to the erosive agent

(*c*) Modifying the landscape to control runoff amounts and rates

(*d*) All the above

Q. 64 Specific practices to avoid water erosion:

(*a*) Growing forage crops in rotation or as permanent cover

(*b*) Growing winter cover crops

(*c*) Inter-seeding

(*d*) All the above

Q. 65 Specific practices to avoid water erosion:

(*a*) Protecting the surface with crop residue

(*b*) Shortening the length and steepness of slopes

(*c*) Increasing water infiltration rates

(*d*) Improving aggregate stability

(*e*) All the above

Q. 66 Specific practices to avoid wind erosion:

(*a*) Maintaining a cover of plants or residue

(*b*) Planting shelterbelts

(*c*) Strip cropping

(*d*) All the above

Q. 67 Specific practices to avoid wind erosion:

(*a*) Increase surface roughness

(*b*) Cultivating on the contour

(*c*) Maintaining soil aggregates at a size less likely to be carried by wind

(*d*) All the above

Q. 68 One of the functions of a wetland is to recharge groundwater sources (True/False):

(*a*) True (*b*) False

Q. 69 **Class III land contains 2% or less slopes (True/False):**

(*a*) True (*b*) False

Q. 70 **Terraces are built and designed around the natural contour of the land (True/False):**

(*a*) True (*b*) False

Q. 71 **Soil erosion is one of the leading offenders in point source pollution (True/False):**

(*a*) True (*b*) False

Q. 72 **One of the main functions of watersheds is controlling floods (True/False):**

(*a*) True (*b*) False

Q. 73 **Leaving adequate ground cover to prevent soil erosion is one of the benefits of maximum tillage (True/False):**

(*a*) True (*b*) False

Q. 74 **A 45 foot strip of grass outlining an agricultural crop-producing field is called a grass waterway (True/False):**

(*a*) True (*b*) False

Q. 75 **One of the benefits of leaving adequate ground cover is that precipitation is allowed to percolate into the soil instead of running off into surrounding surface waters (True/False):**

(*a*) True (*b*) False

Q. 76 **Water quality is affected by the activities within a watershed (True/False):**

(*a*) True (*b*) False

Q. 77 **Agricultural producers convert wetlands into productive acres because it is the least expensive way to increase their production acres (True/False):**

(*a*) True (*b*) False

Q. 78 **Grass waterways are placed in areas of the field where soil erosion is most prevalent (True/False):**

(*a*) True (*b*) False

Q. 79 Which of the following is most likely to cause desertification of soil?

(*a*) Overgrazing of pastures

(*b*) A large accumulation of heavy metals

(*c*) Improper irrigation

(*d*) Improper handling of chemicals

Q. 80 Which type of erosion is the most highly visible?

(*a*) Gully erosion (*b*) Wind erosion

(*c*) Ephemeral gullies (*d*) Splash erosion

Q. 81 The largest of these soil separates or particles is:

(*a*) Gravel (*b*) Sand

(*c*) Silt (*d*) Clay

Q. 82 If soil is stabilized it is:

(*a*) Acted on by geological soil erosion and man-made soil erosion

(*b*) Never moving under any circumstances

(*c*) Under the influence of construction of a new highway

(*d*) Acted on only by geological soil erosion

Q. 83 Which of these is not a key ingredient of soil?

(*a*) Organic matter (*b*) Hemoglobin

(*c*) Minerals (*d*) Air

Q. 84 What percentage of the average soil is organic matter?

(*a*) 45% (*b*) 5%

(*c*) 25% (*d*) 17%

Q. 85 An agronomist is primarily concerned with what use of soil?

(*a*) Construction

(*b*) Acting as a filter for the hydrologic cycle

(*c*) Supporting crop growth

(*d*) Natural beauty

Q. 86 Soil erosion is:

(*a*) The transportation of soil particles from one place to another

(*b*) The result of terraces and diversions

(*c*) Not a natural phenomenon

(*d*) Unable to be affected by humans

Q. 87 Which of the following is not used to reduce soil erosion?

(*a*) Mulch

(*b*) Gravel

(*c*) Residue from previous crops

(*d*) Animal waste and residues

Q. 88 What is no-till crop production?

(*a*) Preparing a good seedbed for crops

(*b*) Ploughing ground, but not discing

(*c*) Planting the seeds in the previous year's crop residue

(*d*) Discing ground prior to planting

Q. 89 Diversion terraces are not suitable for slopes more than:

(*a*) 12% (*b*) 1%

(*c*) 5% (*d*) 7%

Q. 90 When slopes are more than 20% which of the following types of terraces are suitable for:

(*a*) Diversion (*b*) Retention

(*c*) Bench (*d*) All the above

Q. 91 Which of the following is/are the type of waterways?

(*a*) Grassed waterways (*b*) Diversion channels

(*c*) Terrace channels (*d*) All the above

Q. 92 Which of the following is/are (an) example(s) of grass used for grassed waterways?

(*a*) Anjan (*b*) Bermuda

(*c*) Napier (*d*) All the above

Q. 93 The threshold velocity for particle movement in surface wind is:

(*a*) 2km/hr (*b*) 3km/hr

(*c*) 5km/hr (*d*) 10km/hr

Q. 94 Soils containing more than 1.5 MPa soil moisture tension are:

(*a*) Less susceptible for wind erosion

(*b*) More susceptible for wind erosion

(*c*) Both (*a*) and (*b*)

(*d*) None of the above

Q. 95 CSWCRTI was established in which year?

(*a*) 1974 (*b*) 1949

(*c*) 1958 (*d*) 1985

Q. 96 First recorded soil conservation research in India started in which year?

(*a*) 1923 (*b*) 1935

(*c*) 1958 (*d*) 1974

Q. 97 First soil conservation scheme called Dry farming Scheme started at which place in India?

(*a*) Manjiri near Pune (*b*) Dehradun

(*c*) Barapani at Shillong (*d*) None of the above

Q. 98 Individual raindrops strike the soil surface at velocities up to

(*a*) 9m/s (*b*) 5m/s

(*c*) 2m/s (*d*) 1m/s

Q. 99 The first relationship on soil loss to slope length was given by:

(*a*) Zeng

(*b*) Smith

(*c*) Musgrove

(*d*) Wischmeir and Smith

Q. 100 Which of the following soil conservation measure is the most popular in India?

(*a*) Bench terracing (*b*) Contour bunding

(*c*) Contour strip (*d*) All the above

Q. 101 With an increase of R_2O_3 the erodibility of soil:

(*a*) Increase (*b*) Decrease

(*c*) Both (*a*) and (*b*) (*d*) None of the above

Q. 102 Which of the following is/are the types of terrace?

(*a*) Diversion (*b*) Retention

(*c*) Bench (*d*) All the above

Q. 103 When the rate of water supply to the soil surface exceeds the rate of infiltration, the free water, is called:

(*a*) Surface water excess (*b*) Subsurface water

(*c*) Ponding water (*d*) None of the above

Q. 104 Surface storage capacity is the excess water collected:

(*a*) In depressions, (*b*) Forming puddles

(*c*) Both (*a*) and (*b*) (*d*) None of the above

Q. 105 Surface runoff of rain water represents the portion of the water supply to the surface that is:

(*a*) Not absorbed by the soil

(*b*) Not retained on its surface

(*c*) That runs down slope

(*d*) All the above

Q. 106 Overland flow is:

(*a*) Thin Sheet like (*b*) Gully like

(*c*) Both (*a*) and (*b*) (*d*) None of the above

Q. 107 Reynolds number of which value is used as a criterion for the onset of turbulence:

(*a*) 500 (*b*) 200

(*c*) 300 (*d*) 1000

Q. 108 Natural run-off seldom exceeds about what % of annual precipitation:

(*a*) 10 (*b*) 20

(*c*) 30 (*d*) 40

Q. 109 Runoff can be increased several folds by:

(*a*) Stone clearing (*b*) Smoothing

(*c*) Compaction (*d*) All the above

Q. 110 The soil surface can be made water repellent or impermeable by the application of mechanical barriers to water movement such as:

(*a*) Plastic (*b*) Aluminum foil

(*c*) Concrete or sheet metal (*d*) All the above

Q. 111 The soil surface can be made water repellent or impermeable by barriers to water movement involving formation of an impervious soil crust with the aid of various materials, such as:

(*a*) Sodic salts

(*b*) Sprayable asphalt emulsions

(*c*) Silicone water-repellents

(*d*) All the above

Q. 112 Soil erosion is a:

(*a*) Three-stage process (*b*) Two stage process

(*c*) Five stage process (*d*) None of the above

Q. 113 When the runoff water finally comes to rest in a low-lying area, it deposits its suspended load, which is known as:

(*a*) Fluvial sediment (*b*) Sedimentation

(*c*) Both (*a*) and (*b*) (*d*) None of the above

Q. 114 The uniform removal of thin layers of soil from a more or less smooth slope, carried by the distributed (rather than concentrated) flow of runoff water over the soil surface is called:

(*a*) Sheet erosion (*b*) Gully erosion

(*c*) Rill erosion (*d*) None of the above

Q. 115 Splash and sheet erosion are also known as:

(*a*) S inter-rill erosion (*b*) Gully erosion

(*c*) Both (*a*) and (*b*) (*d*) None of the above

Q. 116 The scouring and transport of soil by a concentrated flow of water is called:

(*a*) Rill erosion (*b*) Sheet erosion

(*c*) Both (*a*) and (*b*) (*d*) None of the above

Q. 117 Raindrop sizes range from up to about:

(*a*) 5 mm in diameter (*b*) 10mm in dia

(*c*) 15mm in dia (*d*) 20mm in dia

Q. 118 Raindrops larger than 5 mm because of air turbulence generally tend to:

(*a*) Disintegrate (*b*) Accumulate

(*c*) Both (*a*) and (*b*) (*d*) None of the above

Q. 119 A soil pulverized by excessive tillage or trampling is:

(*a*) More erodible than an untilled and mulch-covered soil.

(*b*) Less erodible than an untilled and mulch-covered soil.

(*c*) Both (*a*) and (*b*)

(*d*) None of the above

Q. 120 When air gusts pick up masses of loose particles from the surface it is called:

(*a*) Deflation (*b*) Inflation

(*c*) Diffraction (*d*) None of the above

Q. 121 The proneness of the soil to wind erosion depends on:

(*a*) Surface dryness

(*b*) Roughness

(*c*) The sizes of particles (texture)

(*d*) Whether the particles are loose

(*e*) Bound in aggregates

(*f*) All the above

Q. 122 The rolling or sweeping motion of relatively large particles that occurs within a few centimeters over the ground surface is called:

(*a*) Surface creep (*b*) Saltation

(*c*) Suspension (*d*) All the above

Q. 123 Surface creep occurs when particle size is:

(*a*) Diameters of 0.5 mm or more

(*b*) Diameter of 10mm or more

(*c*) Diameter of 2mm or more

(*d*) All the above

Q. 124 The jerky movement of finer sand particles that follow distinct trajectories is called:

(*a*) Surface creep (*b*) Saltation

(*c*) Suspension (*d*) All the above

Q. 125 Saltation occurs when particle size is:

(*a*) 0.1–0.5 mm (*b*) 0.5–1 mm

(*c*) 2–5 mm (*d*) 10–20 mm

Q. 126 Methods of land evaluation includes:

(*a*) Qualitative (*b*) Quantitative

(*c*) Both (*a*) and (*b*) (*d*) None of the above

Q. 127 Qualitative method of land evaluation includes:

(*a*) Land capability classification

(*b*) Land irrigability classification

(*c*) Both (*a*) and (*b*)

(*d*) None of the above

Q. 128 Quantitative method of land evaluation includes which of the following:

(*a*) Soil index rating

(*b*) Actual and potential productivity

(*c*) Land suitability classification

(*d*) All the above

Q. 129 Causes of climate change can be:

(*a*) Periodic tidal pulls exerted by astronomical bodies

(*b*) Variation in amount and quality of solar energy (Sun-spot cycle)

(*c*) El Nino/Southern oscillation (ENSO) phenomenon

(*d*) All the above

Q. 130 Positive Effects of climate change can be:

(*a*) Increase of rate of photosynthesis in C_3 plants due high CO_2 conc. and high leaf area

(*b*) High CO_2 conc. reduces transpiration due to reduced stomatal aperture and may result in high water use efficiency

(*c*) Both (*a*) and (*b*)

(*d*) None of the above

Q. 131 Direct Negative Effects of climate change can be:

(*a*) Increased LAI may result in self shading and thus reduces net assimilation rate

(*b*) Increased CO_2 conc. results in higher LAI thus more transpiring surface is available which might compensate increased Water UE

(*c*) Both (*a*) and (*b*)

(*d*) None of the above

Q. 132 Indirect Negative Effects due to increased temperature can be:

(*a*) Reduction in maturity duration of crops with every 1°C rise in temperature

(*b*) Accelerated leaf senescence, forced maturity in winter cereal crops

(*c*) Increased temperature by 1-2 °C above mean temperature of 17 °C during grain filling period reduces yield due to increased rate of senescence of flag leaf and reduction in grain filling period

(*d*) All the above

Q. 133 Indirect Negative Effects due to increased temperature can be:

(a) ET will rise by rise in temperature

(b) Reduction in soil moisture due to increased evaporation from soil and transpiration from plants

(c) Both (a) and (b)

(d) None of the above

Q. 134 The purposes of modeling are:

(a) To reduce a complex system to its essential processes, so that system behaviour may be simulated under different conditions

(b) To reduce time and resources (money, labour, energy)

(c) Both (a) and (b)

(d) None of the above

Q. 135 Different types of models includes:

(a) Mathematical models

(b) Statistical models

(c) Simulation models

(d) All the above

Q. 136 Different types of models includes:

(a) Physical models

(b) Computer models (Information system, decision support system expert system)

(c) Empirical models

(d) All the above

Q. 137 If 50% land is covered by canopy than what can be the probable run-off?

(a) Heavy run-off

(b) Medium run-off

(c) Less run-off

(d) None of the above

Q. 138 If 50-75% land is covered by canopy than what can be the probable run-off?

(a) Heavy run-off

(b) Medium run-off

(c) Less run-off

(d) None of the above

Q. 139 If more than 75% land is covered by canopy than what can be the probable run-off?

(*a*) Heavy run-off (*b*) Medium run-off

(*c*) Less run-off (*d*) None of the above

Q. 140 Of the following, which is the major soil blowing criterion?

(*a*) Soil texture

(*b*) Size of soil aggregates

(*c*) Presence of carbonates

(*d*) Degree of decomposition inorganic soils

Q. 141 Which of the following abbreviations are related with soil blowing grouping?

(*a*) WEG (*b*) EI

(*c*) Both (*a*) and (*b*) (*d*) None of the above

Q. 142 The vulnerability of a soil to erosion, which depends on texture, structure (size and stability of aggregates), looseness and roughness of the soil surface, is called:

(*a*) Erodability (*b*) Susceptibility

(*c*) Vulnerability (*d*) None of the above

Q. 143 The vulnerability of a soil to erosion depends on which of the following properties of soil?

(*a*) Texture

(*b*) Structure

(*c*) Looseness and roughness of the soil surface

(*d*) Size and stability of aggregates

(*e*) All the above

Q. 144 The erosive power of rainfall, running water, or wind is called:

(*a*) Erosivity (*b*) Erodibility

(*c*) Susceptibility (*d*) Run-off

Q. 145 The erosive power of rainfall depends upon:

(*a*) Rainfall intensity (*b*) Duration

(*c*) Both (*d*) None

Q. 146 The erosive power in the case of running water depends on:

(*a*) The velocity (*b*) Turbulence

(*c*) Both (*a*) and (*b*) (*d*) None of the above

Q. 147 An equation relating the height of the water-table between drains to depth of the drains and the distance between them, the hydraulic conductivity of the soil, and the rate of recharge by percolating water is called:

(*a*) Hooghoudt equation (*b*) Darcy equation

(*c*) Renould's equation (*d*) None of the above

Q. 148 The volume of water that runs off the soil surface and flows into streams during episodes of intense rainfall is called:

(*a*) Runoff (*b*) Erosion

(*c*) Splash erosion (*d*) Sheet erosion

Q. 149 The volume of water that runs off during an irrigation that exceeds the soil's infiltrability is called:

(*a*) Run-off (*b*) Erosion

(*c*) Splash erosion (*d*) Sheet erosion

Q. 150 A section of the land surface that drains its surface-water excess to a specified point along a stream is called:

(*a*) Watershed (*b*) Catchment

(*c*) Both (*a*) and (*b*) (*d*) None of the above

Q. 151 Vegetated windbreaks are also called as:

(*a*) Shelterbelts (*b*) Tree belts

(*c*) Watershed (*d*) None of the above

Q. 152 The majority of the land degradation is caused by:

(*a*) Acidification (*b*) Salinization

(*c*) Erosion (*d*) None of the above

Q. 153 Wind and water almost contribute what percentage of soil degradation?

(*a*) 50% (*b*) 60%

(*c*) 70% (*d*) 80%

Q. 154 Canyons are the result of result of:

(*a*) Geological erosion (*b*) Man made erosion

(*c*) Both (*a*) and (*b*) (*d*) None of the above

Q. 155 The maximum amount of soil that can be lost by erosion without causing long term soil degradability is called:

(*a*) T value (*b*) A value

(*c*) E value (*d*) None of the above

Q. 156 T value of a soil depends upon:

(*a*) Organic matter content (*b*) Soil depth

(*c*) Management factors (*d*) All the above

Q. 157 T values generally range from:

(*a*) 5-11Mg/ha (*b*) 1-2Mg/ha

(*c*) 0-1Mg/ha (*d*) None of the above

Q. 158 Water erosion is a:

(*a*) 2 way process (*b*) 1 way process

(*c*) 3 way process (*d*) None of the above

Q. 159 The speed at which the friction between the drop and the air balances the force of gravity is called:

(*a*) Terminal velocity (*b*) Accelerated velocity

(*c*) Threshold velocity (*d*) None of the above

Q. 160 The amount of soil delivered to the stream divided by the amount eroded is called:

(*a*) Delivery ratio (*b*) Terminal velocity

(*c*) Accelerated velocity (*d*) Threshold velocity

Q. 161 Delivery ratio for watersheds where valleys slopes are very steep may reach up to:

(*a*) 20% (*b*) 30%

(*c*) 40% (*d*) 60%

Q. 162 Delivery ratio for small watersheds will be:

(*a*) Less (*b*) More

(*c*) Both (*a*) and (*b*) (*d*) None of the above

Q. 163 Which of the following soil properties influences erodability?

(*a*) Infiltration capacity (*b*) Structural stability

(*c*) Both (*a*) and (*b*) (*d*) None of the above

Q. 164 If the gullies are deep up to 3m and width is not greater than 18m wherein side slopes vary this comes under which class of gully erosion?

(*a*) Very small (*b*) Small

(*c*) Medium (*d*) Deep and narrow

Q. 165 If the gullies are deep up to 3m and width is greater than 18m wherein side slopes vary between 8-15% this comes under which class of gully erosion?

(*a*) Very small (*b*) Small

(*c*) Medium (*d*) Deep and narrow

Q. 166 If the gullies are deep up to 3-9m and width is not less than 18m wherein side slopes vary 8-15% this comes under which class of gully erosion?

(*a*) Very small (*b*) Small

(*c*) Medium (*d*) Deep and narrow

Q. 167 If the gullies are deep >9m and width is variable and slopes mostly steep or even vertical with intricate and active branch gullies, then it is called as:

(*a*) Very small (*b*) Small

(*c*) Medium (*d*) Deep and narrow

Q. 168 The development of gully occurs due to which of the following stages?

(*a*) Formation (*b*) Development

(*c*) Healing (*d*) Stabilization

(*e*) All the above

Q. 169 Agroforestry interventions with adequate soil and moisture conservation methods can reduce the soil loss by:

(*a*) 30% (*b*) 40%

(*c*) 50% (*d*) 70%

Q. 170 No apparent or slight sheet erosion phase is denoted by which symbol?

(*a*) e1 (*b*) e2

(*c*) e3 (*d*) e4

Q. 171 Very severe erosion phase is denoted by which symbol?

(*a*) e1 (*b*) e2

(*c*) e3 (*d*) e4

Q. 172 LCC (Land capability class) 1 is denoted by which colour?

(*a*) Light green (*b*) Yellow

(*c*) Red (*d*) Blue

Q. 173 LCC (Land capability class) 2 is denoted by which colour?

(*a*) Light green (*b*) Yellow

(*c*) Red (*d*) Blue

Q. 174 LCC (Land capability class) 3 is denoted by which colour?

(*a*) Light green (*b*) Yellow

(*c*) Red (*d*) Blue

Q. 175 LCC (Land capability class) 4 is denoted by which colour?

(*a*) Light green (*b*) Yellow

(*c*) Red (*d*) Blue

Q. 176 LCC (Land capability class) 5 is denoted by which colour?

(*a*) Dark green (*b*) Orange

(*c*) Brown (*d*) Purple

Q. 177 LCC (Land capability class) 6 is denoted by which colour?

(*a*) Light green (*b*) Yellow

(*c*) Red (*d*) Blue

Q. 178 LCC (Land capability class) 7 is denoted by which colour?

(*a*) Light green (*b*) Yellow

(*c*) Red (*d*) Blue

Q. 179 LCC (Land capability class) 8 is denoted by which colour?

(*a*) Light green (*b*) Yellow

(*c*) Red (*d*) Blue

Q. 180 If thickness of top soil is 0-15 cm depth the thickness comes under:

(*a*) Less thick (*b*) Moderate thick

(*c*) Thick (*d*) Very thick

Q. 181 If thickness of top soil is 15-30cm depth, the thickness comes under:

(*a*) Less thick (*b*) Moderate thick

(*c*) Thick (*d*) Very thick

Q. 182 If thickness of top soil is 30-60cm depth, the thickness of the top soil comes under:

(*a*) Less thick (*b*) Moderate thick

(*c*) Thick (*d*) Very thick

Q. 183 Hydraulic action of the water will be more pronounced in:

(*a*) Loose soil (*b*) Heavy textured soil

(*c*) Both (*a*) and (*b*) (*d*) None of the above

Q. 184 The particles of the running water collide with each other resulting in to further breakdown of the particles This is called:

(*a*) Attrition (*b*) Abrasion

(*c*) Solution (*d*) None of th ɔove

Q. 185 Which of the following erosion can be easily re ved by normal tillage operations?

(*a*) Rill erosion (*b*) Gully erosi

(*c*) Sheet erosion (*d*) Splash eros

Q. 186 Most common size of raindrop is:

(*a*) 2mm (*b*) 3mm

(*c*) 4mm (*d*) 7mm

Q. 187 The maximum impact of the raindrop will be at what angle when it hits the ground?

(*a*) 30° (*b*) 45°

(*c*) 90° (*d*) 180°

Q. 188 Which of the following soil separates has more erodability?

(*a*) Sand (*b*) Silt

(*c*) Clay (*d*) Rocks

Q. 189 Which of the following types of clay minerals will provide more stable aggregates?

(*a*) Nonexpanding type of clay minerals

(*b*) Expanding type of clay minerals

(*c*) Both (*a*) and (*b*)

(*d*) None of the above

Q. 190 The erodability increase at what size of aggregates?

(*a*) 0.2mm (*b*) 0.3mm

(*c*) 0.4mm (*d*) 0.5mm

Q. 191 Which is the most important index of soil erodibility?

(*a*) Silica content (*b*) Sesquioxide content

(*c*) Silica-sesquioxide ratio (*d*) All the above

Q. 192 Match the following:

a. Man-made erosion	**i. Vegetative erosion control**
b. Splash erosion	**ii. Very productive, nearly level**
c. Class-I	**iii. Topsoil being washed off of cultivated fields**
d. Cover crops	**iv. Rain drops shatters surface aggregates**

(*a*) i, ii, iii, iv (*b*) iii, iv, ii, i

(*c*) ii, iii, i, iv (*d*) iv, iii, ii, i

Q. 193 Match the following:

a. Geological erosion	**i. Severe limitations**
b. Terraces	**ii. Wearing away of land's surface by water, wind, and ice**
c. Soil erosion	**iii. Wearing away of earth's surface under natural environmental conditions**
d. Class IV	**iv. Mechanical erosion control**

(*a*) i, ii, iii, iv (*b*) iii, iv, ii, i

(*c*) ii, iii, i, iv (*d*) iv, iii, ii, i

Q. 194 The major role of inter-cultivation tillage is:

(*a*) Improvement of soil texture

(*b*) Enhancement of soil aeration

(*c*) Weed control

(*d*) Improvement of soil structure

ANSWERS

1	(*c*)	23	(*a*)	45	(*b*)	67	(*d*)
2	(*a*)	24	(*c*)	46	(*d*)	68	(*a*)
3	(*b*)	25	(*b*)	47	(*d*)	69	(*b*)
4	(*d*)	26	(*d*)	48	(*a*)	70	(*a*)
5	(*b*)	27	(*c*)	49	(*a*)	71	(*b*)
6	(*a*)	28	(*a*)	50	(*d*)	72	(*b*)
7	(*a*)	29	(*c*)	51	(*d*)	73	(*b*)
8	(*d*)	30	(*b*)	52	(*d*)	74	(*b*)
9	(*c*)	31	(*c*)	53	(*c*)	75	(*a*)
10	(*d*)	32	(*a*)	54	(*d*)	76	(*a*)
11	(*d*)	33	(*b*)	55	(*e*)	77	(*a*)
12	(*c*)	34	(*c*)	56	(*e*)	78	(*a*)
13	(*b*)	35	(*c*)	57	(*c*)	79	(*a*)
14	(*c*)	36	(*c*)	58	(*e*)	80	(*a*)
15	(*a*)	37	(*b*)	59	(*d*)	81	(*a*)
16	(*c*)	38	(*b*)	60	(*f*)	82	(*d*)
17	(*d*)	39	(*d*)	61	(*d*)	83	(*b*)
18	(*b*)	40	(*d*)	62	(*e*)	84	(*b*)
19	(*d*)	41	(*a*)	63	(*d*)	85	(*c*)
20	(*a*)	42	(*d*)	64	(*d*)	86	(*a*)
21	(*a*)	43	(*a*)	65	(*e*)	87	(*d*)
22	(*d*)	44	(*e*)	66	(*d*)	88	(*c*)

89	(*a*)	116	(*a*)	143	(*e*)	170	(*a*)
90	(*c*)	117	(*a*)	144	(*a*)	171	(*d*)
91	(*d*)	118	(*a*)	145	(*c*)	172	(*a*)
92	(*d*)	119	(*a*)	146	(*c*)	173	(*b*)
93	(*b*)	120	(*a*)	147	(*a*)	174	(*c*)
94	(*a*)	121	(*f*)	148	(*a*)	175	(*d*)
95	(*a*)	122	(*a*)	149	(*a*)	176	(*a*)
96	(*a*)	123	(*a*)	150	(*c*)	177	(*b*)
97	(*a*)	124	(*b*)	151	(*a*)	178	(*c*)
98	(*a*)	125	(*a*)	152	(*c*)	179	(*d*)
99	(*a*)	126	(*c*)	153	(*d*)	180	(*a*)
100	(*b*)	127	(*c*)	154	(*a*)	181	(*b*)
101	(*a*)	128	(*d*)	155	(*a*)	182	(*c*)
102	(*d*)	129	(*d*)	156	(*d*)	183	(*d*)
103	(*a*)	130	(*c*)	157	(*a*)	184	(*a*)
104	(*c*)	131	(*c*)	158	(*c*)	185	(*a*)
105	(*d*)	132	(*d*)	159	(*a*)	186	(*d*)
106	(*a*)	133	(*c*)	160	(*a*)	187	(*c*)
107	(*a*)	134	(*c*)	161	(*d*)	188	(*b*)
108	(*a*)	135	(*d*)	162	(*b*)	189	(*a*)
109	(*d*)	136	(*d*)	163	(*c*)	190	(*a*)
110	(*d*)	137	(*a*)	164	(*a*)	191	(*d*)
111	(*d*)	138	(*b*)	165	(*b*)	192	(*b*)
112	(*a*)	139	(*a*)	166	(*c*)	193	(*b*)
113	(*c*)	140	(*d*)	167	(*d*)	194	(*d*)
114	(*a*)	141	(*c*)	168	(*e*)		
115	(*a*)	142	(*a*)	169	(*d*)		

Chapter 11

Memory Based ARS/SRF Questions

Q. 1 Sheet erosion is observed on:

(*a*) Deeply sloping land

(*b*) Moderately sloping land

(*c*) Slopy land

(*d*) Gently sloping land

Q. 2 Run-off and soil loss can be calculated by:

(*a*) Ramser's formula

(*b*) Chepil equation

(*c*) Chepil and Woodruff equation

(*d*) All

Q. 3 The quantity of soil materials of a given size in the run-off varies as the ______ power of the speed of run-off:

(*a*) Third (*b*) Fourth

(*c*) Fifth (*d*) Sixth

Q. 4 Erosion permitting crops is/are:

(*a*) Maize (*b*) Wheat

(*c*) Bajra (*d*) All

Q. 5 Soils more susceptible to water erosion are:

(*a*) Coarse textured soils

(*b*) Medium textured soils

(*c*) Fine textured soils

(*d*) All

Q. 6 In Universal soil loss equation, A=RKLSCP, K denotes:

(*a*) Soil conservation practices (*b*) Soil roughness

(*c*) Soil erodability (*d*) Rainfall factor

Q. 7 Minimum wind speed required to initiate the movement of most erodable soil particles is:

(*a*) 10 km (*b*) 12 km

(*c*) 14 km (*d*) 16 km

Q. 8 Diameter of particles that is most susceptible to wind erosion is:

(*a*) 0.1 mm (*b*) 0.2 mm

(*c*) 0.02 mm (*d*) 0.01 mm

Q. 9 Dominant type of wind erosion in soil is:

(*a*) Saltation (*b*) Surface creep

(*c*) Suspension (*d*) All

Q. 10 Size range of soil particles in saltation type of wind erosion is:

(*a*) Less than 0.05 mm (*b*) 0.5–1.0 mm

(*c*) 0.5–2.0 mm (*d*) 0.05–0.5 mm

Q. 11 Size range of soil particles in suspension type of wind erosion is:

(*a*) Less than 0.05 mm (*b*) 0.5–1.0 mm

(*c*) 0.5–2.0 mm (*d*) 0.05–0.5 mm

Q. 12 Size range of soil particles in surface creep type of wind erosion is:

(*a*) Less than 0.05 mm (*b*) 0.5–1.0 mm

(*c*) 0.5–2.0 mm (*d*) 0.05–0.5 mm

Q. 13 In Chepil and Woodruff's equation of soil loss calculation by wind erosion, E = f (ICKLV), I denotes:

(*a*) Soil erodability (*b*) Rainfall intensity

(*c*) Soil roughness (*d*) Soil type

Q. 30 What is the unit of gravity?

(a) cm sec^{-1}
(b) g cm^{-3}
(c) g sec cm^{-1}
(d) cm sec^{-2}

Q. 31 In Stoke's law 'η' denotes:

(a) Density
(b) Gravity
(c) Viscosity
(d) Velocity

Q. 32 Textural analysis of all soils except calcareous and organic soils can be done by:

(a) Hydrometer method
(b) International pipette method
(c) Aggregate method
(d) All

Q. 33 More accurate analysis of soil texture can be done by:

(a) Hydrometer method
(b) International pipette method
(c) Aggregate method
(d) All

Q. 34 The arrangement of primary and secondary soil particles is called as:

(a) Soil structure
(b) Soil texture
(c) Soil consistency
(d) Soil plasticity

Q. 35 The relative proportion of soil particles is called as:

(a) Soil structure
(b) Soil texture
(c) Soil consistency
(d) Soil plasticity

Q. 36 Wet bulk density (g cm^{-3}):

(a) $\frac{\text{Weight of soil}}{\text{Volume of soil}}$

(b) $\frac{\text{Weight of dry soil}}{\text{Volume of soil}}$

(c) $\dfrac{\text{Weight of dry soil}}{\text{Volume of soil particles}}$

(d) $\dfrac{\text{Volume of soil}}{\text{Weight of moist soil}}$

Q. 37 Percentage of pore space:

(a) [1–Bulk density/Particle density] × 100
(b) [10–Bulk density/Particle density] × 100
(c) [1–Particle density/Bulk density] × 100
(d) [10–Particle density/Bulk density] × 100

Q. 38 Volume wetness = Mass wetness × ________:

(a) Bulk density (b) Particle density
(c) Weight of soil (d) Pore space

Q. 39 What is the depth of plough layer of soil?

(a) 0–10 cm (b) 0–15 cm
(c) 0–20 cm (d) 0–25 cm

Q. 40 Element essential for the flocculation of soil particles is:

(a) Sodium (b) Potassium
(c) Calcium (d) Magnesium

Q. 41 Element essential for the deflocculation of soil particles is:

(a) Potassium (b) Calcium
(c) Magnesium (d) Sodium

Q. 42 Zeta potential of soil particles is reduced by:

(a) Potassium (b) Calcium
(c) Sodium (d) Magnesium

Q. 43 Increase in zeta potential of soil particles leads to:

(a) Flocculation (b) Deflocculation
(c) Aggregation (d) Both (a) and (b)
(e) Both (a) and (c)

Q. 44 Cementation of soil particles may be caused by:

(a) Cations

(b) Oxides of iron and aluminium

(c) Anions

(d) Humus substances

(e) All

Q. 45 Indirect methods of soil structure characterization and evaluation include:

(a) Wet and dry sieving methods

(b) Mean weight diameter and GMD

(c) Stability index and structural co-efficient

(d) All the above

Q. 46 The method of soil structure characterization and evaluation that was described by Chepil is:

(a) Stability index

(b) Wet sieving

(c) Dry sieving

(d) Mean weight diameter

Q. 47 The method of soil structure characterization and evaluation that was described by Yodder is:

(a) Stability index

(b) Wet sieving

(c) Dry sieving

(d) Mean weight diameter

Q. 48 For better soil structure, the value of cumulative mean weight diameter should be:

(a) Lower (b) Higher

(c) Unity (d) Zero

Q. 49 For better soil structure, the value of structural co-efficient should be:

(a) Lower (b) Higher

(c) Unity (d) Zero

Q. 50 The physical condition of soil in relation to plant growth is called as:

(*a*) Soil tilth (*b*) Tillage

(*c*) Coherence (*d*) Adherence

Q. 51 The distance traveled by water through soil column is called as:

(*a*) Infiltration rate (*b*) Percolation rate

(*c*) Seepage rate (*d*) Run-off loss

Q. 52 The most important factor(s) that influences the infiltration of water is/are:

(*a*) Soil structure (*b*) Soil texture

(*c*) Porosity (*d*) All

Q. 53 Infiltration rate is influenced by:

(*a*) Soil texture

(*b*) Soil structure

(*c*) Porosity

(*d*) Initial soil water content

(*e*) All the above

Q. 54 Rocks formed by cooling of the molten materials (magma) from the sun is called as:

(*a*) Metamorphic rocks (*b*) Igneous rocks

(*c*) Sedimentary rocks (*d*) All

Q. 55 Of the followings, which is/are the examples of igneous rocks?

(*a*) Granites (*b*) Basalts

(*c*) Syenites (*d*) All

Q. 56 Rocks formed from igneous rocks by the influence of high pressure and temperature are called as:

(*a*) Metamorphic rocks (*b*) Primary rocks

(*c*) Sedimentary rocks (*d*) All

Q. 57 Gneiss, Schist and Slate are the examples of:

(*a*) Metamorphic rocks (*b*) Sedimentary rocks

(*c*) Primary rocks (*d*) Igneous rocks

Q. 58 Which is/are the chemical weathering processes?

(*a*) Carbonation (*b*) Hydrolysis

(*d*) Hydration (*d*) All

Q. 59 Which of the following processes is responsible for the formation of serpentine from olivine?

(*a*) Hydrolysis (*b*) Hydration

(*c*) Oxidation (*d*) Reduction

Q. 60 Most easily weatherable mineral is:

(*a*) Gypsum (*b*) Biotite

(*c*) Olivine (*d*) Calcite

Q. 61 Mineral formed by exfoliation, a type of physical weathering, is:

(*a*) Granite (*b*) Basalt

(*c*) Gypsum (*d*) Calcite

Q. 62 Stability of minerals to weathering and to decomposition is related to:

(*a*) Degree of basicity

(*b*) Degree of linkage to tetrahedron

(*c*) Silica: Sesquioxide ratio

(*d*) All

Q. 63 When bases are more, the stability of minerals will:

(*a*) Decrease (*b*) Increase

(*c*) Unaltered (*d*) None

Q. 64 Clay minerals predominate in the soils of arid regions are:

(*a*) Montmorillonite and vermiculite

(*b*) Illite and montmorillonite

(*c*) Illite and chlorite

(*d*) Illite and vermiculite

Q. 65 The natural aggregation of soil separates into peds is called as:

(*a*) Soil textural class (*b*) Soil separates

(*c*) Soil structure (*d*) Soil consistency

Q. 66 If 100 cm^3 of soil has an oven dry weight of 135 grams, what is its bulk density?

(*a*) 13,500 g/cm^3 (*b*) 0.75 cm^3/g

(*c*) 1.35 g/cm^3 (*d*) 1.53 g/cm^3

Q. 67 Soil bulk density is greatest for:

(*a*) Sand (*b*) Clay

(*c*) Peat (*d*) Muck

Q. 68 Soil compaction results in (an) ________ in soil bulk density:

(*a*) Increase (*b*) Decrease

(*c*) No change (*d*) None

Q. 69 Insoluble fraction of organic matter is:

(*a*) Humin (*b*) Fulvic acid

(*c*) Humic acid (*d*) Hymatomelanic acid

Q. 70 Organic compound that contain more oxygen but less carbon and nitrogen is:

(*a*) Humic acid

(*b*) Fulvic acid

(*c*) Humin

(*d*) Hymatomelanic acid

Q. 71 The components of Atomic Absorption Spectrophotometer Instrument are:

(*a*) A detector that measures the amount of absorption

(*b*) Electronic amplification of the absorption signal;

(*c*) A readout system that normally is a strip chart recorder, a digital display, a meter, or a printer

(*d*) All the above

Q. 72 Principle of operation in Atomic Absorption Spectrophotometer instruments is:

(*a*) The sample aerosol is aspirated

(*b*) Mixed with the gases before combustion

(*c*) During combustion, the element is reduced to the atomic state and then can absorb radiation

(*d*) Absorption of radiation is directly proportional to no. of atoms present in ground state

(*e*) All the above

Q. 73 Arrange the following minerals based on their stability in increasing order:

(*a*) Albite > Zircon > Magnetite > Muscovite

(*b*) Magnetite > Muscovite > Albite > Zircon

(*c*) Magnetite > Zircon > Albite > Muscovite

(*d*) Albite > Muscovite > Magnetite > Zircon

Q. 74 Which is an amorphous clay mineral?

(*a*) Kaolinite (*b*) Smectite

(*c*) Allophane (*d*) Chlorite

Q. 75 Of the following, which is/are example(s) for 1:1 dioctahedral clay minerals?

(*a*) Serpentine (*b*) Kaolinite

(*c*) Halloysite (*d*) Both (*a*) and (*b*)

(*e*) Both (*b*) and (*c*)

Q. 76 Of the following, which is/are example(s) for 1:1 trioctahedral clay minerals?

(*a*) Greenalite (*b*) Serpentine

(*c*) Chrysolite (*d*) All

Q. 77 What is the basal spacing of montmorillonite?

(*a*) 10 Å (*b*) 10.25 Å

(*c*) 7.2 Å (*d*) 14 Å

Q. 78 _____________ **is an example for 2:1 type of non-expanding clay minerals:**

(a) Mica (b) Smectite

(c) Chlorine (d) Kaoline

Q. 79 What is the basal spacing of illite or hydrous mica?

(a) 7.2 Å (b) 14 Å

(c) 10 Å (d) 10.25 Å

Q. 80 The 2: 1 non-expanding clay mineral _____________ **exists both in dioctahedral and trioctahedral forms:**

(a) Chlorite (b) Glauconite

(c) Vermiculite (d) Illite

Q. 81 What is the total surface area of montmorillonite (m^2 kg^{-1})?

(a) 580–750 × 10^3 (b) 130–180 × 10^3

(c) 780–900 × 10^3 (d) 120–170 × 10^3

Q. 82 What is the external surface area of montmorillonite (m^2 kg^{-1})?

(a) 50–70 × 10^3 (b) 70–100 × 10^3

(c) 80–100 × 10^3 (d) 80–150 × 10^3

Q. 83 Total surface area of chlorite (m^2 kg^{-1}) is:

(a) 37–45 × 10^3 (b) 130–180 × 10^3

(c) 120–170 × 10^3 (d) 580–750 × 10^3

Q. 84 CEC of montmorillonite clay mineral (cmol (p+) kg^{-1}) is:

(a) 69–81 (b) 57–64

(c) 80–150 (d) 30–40

Q. 85 What is the CEC of kaolinite (cmol (p+) kg^{-1})?

(a) 30–40 (b) 3–15

(c) 18–22 (d) 57–64

Q. 86 Which of the following sensors was/were used in IRS-1C

(a) PAN (b) LISS

(c) WiFS (d) All

Q. 87 Which of the following satellite is the most recent one:

(*a*) IRS-1D (*b*) IRS-P2

(*c*) Oceansat-1 (*d*) Risat

Q. 88 What is the distance of sun-synchronous satellite from the earth surface:

(*a*) 600–1000 km (*b*) 1000–2000 km

(*b*) 2000–3000 km (*d*) 36000 km

Q. 89 The point on the ground vertically beneath the respective centre of the camera lens is known as:

(*a*) NADIR (*b*) Zenith

(*c*) Both (*d*) None

Q. 90 How many satellites are existing in one orbit of a GPS system:

(*a*) 2 (*b*) 4

(*c*) 6 (*d*) 8

Q. 91 GPS satellites are fixed at a distance of:

(*a*) 20,200km (*b*) 10,200km

(*c*) 900km (*d*) 600km

Q. 92 The orbital period of each GPS satellite is:

(*a*) 6 hrs (*b*) 4hrs

(*c*) 12hrs (*d*) 10hrs

Q. 93 NAVSTAR belongs to which country:

(*a*) India (*b*) Russia

(*c*) USA (*d*) China

Q. 94 Each GPS satellite is provided with how many accurate clocks:

(*a*) 1 (*b*) 2

(*c*) 3 (*d*) 4

Q. 95 L1 frequency of GPS receiver works on:

(*a*) 1575.4 MHtz (*b*) 1227.6 MHtz

(*c*) Both (*d*) None

Q. 96 L2 frequency of GPS receiver works on:

(*a*) 1575.4 MHtz (*b*) 1227.6 MHtz

(*c*) Both (*d*) None

Q. 97 L1 frequency modulated with which of the following codes:

(*a*) P (*b*) C/A

(*c*) Both (*a*) and (*b*) (*d*) None

Q. 98 Energy of hydrogen bond in H-OH is ....

(*a*) 0.72 Kcal/mole (*b*) 7.2 Kcal/mole

(*c*) 72 Kcal/mole (*d*) 720 Kcal/mole

Q. 99 The bond length of H-H is:

(*a*) 0.74 Å (*b*) 2.72 Å

(*c*) 3.23 Å (*d*) 2.78 Å

Q. 100 Dipole moment of each O-H bond of H_2O is:

(*a*) 1.60 D (*b*) 1.85 D

(*c*) 2.60 D (*d*) 285 D

ANSWERS

1	(*d*)	23	(*a*)	45	(*d*)	67	(*a*)
2	(*a*)	24	(*d*)	46	(*c*)	68	(*a*)
3	(*c*)	25	(*a*)	47	(*b*)	69	(*a*)
4	(*d*)	26	(*d*)	48	(*a*)	70	(*b*)
5	(*c*)	27	(*b*)	49	(*b*)	71	(*d*)
6	(*c*)	28	(*c*)	50	(*a*)	72	(*e*)
7	(*d*)	29	(*c*)	51	(*a*)	73	(*d*)
8	(*a*)	30	(*d*)	52	(*c*)	74	(*c*)
9	(*a*)	31	(*c*)	53	(*e*)	75	(*e*)
10	(*d*)	32	(*a*)	54	(*b*)	76	(*d*)
11	(*a*)	33	(*b*)	55	(*d*)	77	(*d*)
12	(*c*)	34	(*a*)	56	(*c*)	78	(*a*)
13	(*a*)	35	(*b*)	57	(*a*)	79	(*c*)
14	(*d*)	36	(*a*)	58	(*d*)	80	(*c*)
15	(*a*)	37	(*c*)	59	(*b*)	81	(*a*)
16	(*c*)	38	(*a*)	60	(*c*)	82	(*d*)
17	(*a*)	39	(*b*)	61	(*a*)	83	(*b*)
18	(*a*)	40	(*c*)	62	(*d*)	84	(*c*)
19	(*a*)	41	(*d*)	63	(*b*)	85	(*b*)
20	(*a*)	42	(*b*)	64	(*c*)	86	(*d*)
21	(*b*)	43	(*b*)	65	(*c*)	87	(*d*)
22	(*c*)	44	(*e*)	66	(*c*)	88	(*a*)

89	**(*a*)**	**92**	**(*c*)**	**95**	**(*a*)**	**98**	**(*b*)**
90	**(*b*)**	**93**	**(*c*)**	**96**	**(*b*)**	**99**	**(*a*)**
91	**(*a*)**	**94**	**(*d*)**	**97**	**(*c*)**	**100**	**(*a*)**

Chapter 12

Exercises

Q. 1 Surface area of spheroid can be calculated by which of the following formula?

(*a*) πr^3 (*b*) $4/3\ \pi r^3$

(*c*) $2\ \pi r$ (*d*) $\pi r^2 h$

Q. 2 Particle smaller than _______ microns cannot be separated by sieving:

(*a*) <2μ (*b*) < 20μ

(*c*) < 50μ (*d*) < 100μ

Q. 3 What is the unit of viscosity?

(*a*) cm sec^{-1} (*b*) g cm^{-3}

(*c*) g sec cm^{-1} (*d*) cm sec^{-2}

Q. 4 What is the unit of gravity?

(*a*) cm sec^{-1} (*b*) g cm^{-3}

(*c*) g sec cm^{-1} (*d*) cm sec^{-2}

Q. 5 The equation, $v = K.r^2$ is called as:

(*a*) Stoke's law (*b*) Ratio law

(*c*) Liebig law (*d*) Mitcherlich equation

Q. 6 In Stoke's law 'η' denotes:

(*a*) Density (*b*) Gravity

(*c*) Viscosity (*d*) Velocity

Q. 7 What is the value of particle density in Stoke's law?

(*a*) 1.1–1.8 g cm^{-3} (*b*) 2.25 g cm^{-3}

(*c*) 2.45 g cm^{-3} (*d*) 2.65 g cm^{-3}

Q. 8 Soils that have 60 % clay, 20 % sand and 20 % silt can be grouped in to the textural class of:

(*a*) Clay (*b*) Silty clay

(*c*) Silt (*d*) Silty clay loam

Q. 9 Plasticity and stickiness are high in:

(*a*) Sandy soils (*b*) Loamy soils

(*c*) Clay soils (*d*) All

Q. 10 Theory of bonding of clay-clay particles was given by:

(*a*) E.J. Russell (*b*) R.D. Jenny

(*c*) Juston von Leibig (*d*) Waag and Guldberg

Q. 11 Anion responsible for soil aggregation is:

(*a*) Nitrate (*b*) Phosphate

(*c*) Sulphate (*d*) Molybdate

Q. 12 Hydroxides of iron and aluminium on ________ may form stable aggregates in laterite soils:

(*a*) Oxidation (*b*) Reduction

(*c*) Hydration (*d*) Dehydration

Q. 13 More stable aggregates are formed by:

(*a*) Humus substances (*b*) Earthworms

(*c*) Plant roots decay (*d*) All

Q. 14 Total head drop ($\Delta H/L$) per unit distance in the soil water column is called as:

(*a*) Flux (*b*) Flux density

(*c*) Hydraulic conductivity (*d*) Hydraulic gradient

Q. 15 Rate of flow of liquid through a porous medium under unit hydraulic gradient is called as:

(*a*) Darcy's law (*b*) Flux density

(*c*) Hydraulic conductivity (*d*) Stoke's law

Q. 16 The formula $q = k\Delta H/L$ is:

(*a*) Darcy's law (*b*) Hydraulic gradient

(*c*) Flux density (*d*) Fick's law

Q. 17 Hydraulic conductivity of soil is found to increase in which of the following order?

(*a*) Na–soil > K–soil > Ca–soil

(*b*) Ca–soil > K–soil > Na–soil

(*c*) Ca–soil > Na–soil > K- soil

(*d*) Na–soil > Ca–soil > K–soil

Q. 18 Permeability classification of soil was given by:

(*a*) O 'Neal' (*b*) Atterberg

(*c*) Schofield (*d*) Chepil and Woodruff

Q. 19 What is the total permeability class of soil given by O 'Neal'?

(*a*) Four (*b*) Five

(*c*) Six (*d*) Seven

Q. 20 What is the unit of hydraulic conductivity?

(*a*) cm sec^{-2} (*b*) m sec^{-1}

(*c*) cm sec^{-1} (*d*) m sec^{-2}

Q. 21 In the laboratory, hydraulic conductivity is measured by:

(*a*) Piezometer

(*b*) Constant head permeameter

(*c*) Auger hole method

(*d*) Both (*a*) and (*b*)

Q. 22 What is the source of gamma radiation in gamma-ray scanner used for soil water estimation?

(*a*) Cesium 137 (*b*) Strontium

(*c*) Lanthanum (*d*) Silver

Q. 23 Thermal conductivity follows the order of

(*a*) Peat > Clay > Loam > Sand

(*b*) Sand > Loam > Clay > Peat

(*c*) Loam > Peat > Clay > Sand

(*d*) Peat > Loam > Clay > Sand

Q. 24 Bulk density of organic matter is:

(*a*) 0.3 g/cc (*b*) 0.5 g/cc

(*c*) 0.6 g/cc (*d*) 1.0 g/cc

Q. 25 What is unit of cation exchange capacity?

(*a*) meq/100g (*b*) cmol(p+)/kg

(*c*) mmho/100g (*d*) All

Q. 26 What is the SI unit of cation exchange capacity?

(*a*) meq/100g (*b*) cmol(p+)/ kg

(*c*) mmho/100g (*d*) All

Q. 27 Equivalency, reversibility and stoichiometry are the characteristics of which of the following?

(*a*) Anion exchange capacity

(*b*) Cation exchange capacity

(*c*) Both (*a*) and (*b*)

(*d*) None

Q. 28 High Pressure Liquid Chromatography (HPLC) involves which of the following steps:

(*a*) Separation of metal species

(*b*) Formation coloured complex

(*c*) Determination through absorbance detector

(*d*) All the above

Q. 29 What is/are the basic types of HPLC instrument systems which are available?

(*a*) Isocratic (*b*) Gradient

(*c*) Both (*a*) and (*b*) (*d*) None of the above

Q. 30 Which of the following is more stable mineral?

(*a*) Zircon (*b*) Tourmaline

(*c*) Quartz (*d*) Magnetite

Q. 31 Which of the following is easily weatherable mineral?

(*a*) Olivine (*b*) Hornblende

(*c*) Augite (*d*) Biotite

Q. 32 Arrange the following minerals based on their stability in increasing order.

(*a*) Albite > Zircon > Magnetite > Muscovite

(*b*) Magnetite > Muscovite > Albite > Zircon

(*c*) Magnetite > Zircon > Albite > Muscovite

(*d*) Albite > Muscovite > Magnetite > Zircon

Q. 33 Arrange the following minerals based on their stability in decreasing order:

(*a*) Biotite >Apatite > Anorthite > Olivine

(*b*) Anorthite > Apatite > Biotite > Olivine

(*c*) Anorthite > Biotite > Apatite > Olivine

(*d*) Biotite > Anorthite > Apatite > Olivine

Q. 34 Which one of the following is the light mineral?

(*a*) Quartz (*b*) Feldspar

(*c*) Mica (*d*) Microcline

(*e*) All the above

Q. 35 Mineral which has more internal surface area is:

(*a*) Vermiculite (*b*) Illite

(*c*) Montmorillonite (*d*) Kaolinite

Q. 36 Mineral which has more external surface area is:

(*a*) Vermiculite (*b*) Illite

(*c*) Montmorillonite (*d*) Kaolinite

Q. 37 What is the total surface area of montmorillonite ($m^2\ kg^{-1}$)?

(*a*) 580–750 × 10^3 (*b*) 130–180 × 10^3

(*c*) 780–900 × 10^3 (*d*) 120–170 × 10^3

Q. 38 What is the external surface area of montmorillonite ($m^2\ kg^{-1}$)?

(*a*) 50–70 × 10^3 (*b*) 70–100 × 10^3

(*c*) 80–100 × 10^3 (*d*) 80–150 × 10^3

Q. 39 Total surface area of chlorite ($m^2\ kg^{-1}$) is:

(*a*) 37–45 × 10^3 (*b*) 130–180 × 10^3

(*c*) 120–170 × 10^3 (*d*) 580–750 × 10^3

Q. 40 CEC of montmorillonite clay mineral (cmol (p+) kg^{-1}) is:

(*a*) 69–81 (*b*) 57–64

(*c*) 80–150 (*d*) 30–40

Q. 41 What is the CEC of kaolinite (cmol (p+) kg^{-1})?

(*a*) 30–40 (*b*) 3–15

(*c*) 18–22 (*d*) 57–64

Q. 42 Clay minerals can be identified by:

(*a*) X-ray Diffraction Analysis

(*b*) Electron Diffraction Analysis

(*c*) Differential Thermal Analysis

(*d*) All

Q. 43 What Inert material is used in Differential Thermal Analysis for clay minerals identification?

(*a*) Ceramic oxide (*b*) Hydrogen peroxide

(*c*) Aluminium oxide (*d*) Ethylene glycol

Q. 44 Crystallization of clay minerals is a ____________ process:

(*a*) Exothermic (*b*) Endothermic

(*c*) Both (*a*) and (*b*) (*d*) Inert

Q. 45 Which of the following is/are the application of computer technology?

(*a*) Soil mapping at cadastral scale using high resolution spatial, spectral and radiometric resolutions

(*b*) Quantification of soil loss

(*c*) Detection of water logging due to rising ground water table

(*d*) All the above

Q. 46 Which of the following is/are the application of computer technology?

(*a*) Delineation of salt-affected soils in black soil and sandy regions

(*b*) Soil moisture estimation and mapping using micro wave/optical/thermal remote sensing techniques in surface and root zone depth

(*c*) Land surface temperature estimation using thermal and microwave remote sensing techniques

(*d*) All the above

Q. 47 Which of the following is/are the application of computer technology?

(*a*) Hyper spectral studies on soils to establish quantitative relationship between spectral reflectance and soil properties

(*b*) Development of digital techniques for a variety of applications using GIS techniques for e.g., soil suitability to crops, land capability classification and land irrigability assessment etc.

(*c*) Yield models by integration of biophysical simulation and regional level crop models

(*d*) All the above

Q. 48 Which of the following is/are the application of computer technology?

(*a*) Land information system (*b*) Market intelligent

(*c*) Pest management system (*d*) All the above

49 Which of the following can be a suitable area for computer application?

(*a*) Identification of crops and estimation of area and production of short duration crops grown in fragmented land holdings

(*b*) Forecasting of droughts/floods

(*c*) Detection of crop stress due to nutrients, pests and diseases and quantification of their effects on crop yield (*d*) Automation of land evaluation procedures for a variety of applications using GIS techniques

(*e*) All the above

Q. 50 Which of the following can be a suitable area for computer application?

(*a*) Information on sub-surface soil horizons

(*b*) Extending precision farming database to smaller farm size and/or diverse crops/cropping systems

(*c*) Developing decision support systems for management of biotic and abiotic stresses at the farm level

(*d*) More accurate yield models

(*e*) All the above

Q. 51 Strong dusty winds are locally called as:

(*a*) Aeolian (*b*) Loo

(*c*) Loess (*d*) None

Q. 52 Minimum wind speed required to initiate the movement of most erodable soil particles is:

(*a*) 10 km (*b*) 12 km

(*c*) 14 km (*d*) 16 km

Q. 53 Diameter of particles that is most susceptible to wind erosion is:

(*a*) 0.1 mm (*b*) 0.2 mm

(*c*) 0.02 mm (*d*) 0.01 mm

Q. 54 Dominant type of wind erosion in soil is:

(*a*) Saltation (*b*) Surface creep

(*c*) Suspension (*d*) All

Q. 55 What are signs of water erosion?

(*a*) Pedestals of soil supporting pebbles and plant material

(*b*) Convex landscape positions

(*c*) Both (*a*) and (*b*)

(*d*) None of the above

Q. 56 Long-term soil erosion results in:

(*a*) Persistent and large gullies

(*b*) Exposure of lighter coloured subsoil at the surface

(*c*) Poorer plant growth

(*d*) All the above

Q. 57 The soil surface can be made water repellent or impermeable by the application of mechanical barriers to water movement such as:

(*a*) Plastic (*b*) Aluminum foil

(*c*) Concrete or sheet metal (*d*) All the above

Q. 58 The soil surface can be made water repellent or impermeable by barriers to water movement involving formation of an impervious soil crust with the aid of various materials, such as:

(*a*) Sodic salts

(*b*) Sprayable asphalt emulsions

(*c*) Silicone water-repellents

(*d*) All the above

Q. 59 Soil erosion is a:

(*a*) Three-stage process (*b*) Two-stage process

(*c*) Five-stage process (*d*) None of the above

Q. 60 When the runoff water finally comes to rest in a low-lying area, it deposits its suspended load, which is known as:

(*a*) Fluvial sediment (*b*) Sedimentation

(*c*) Both (*a*) and (*b*) (*d*) None of the above

Q. 61. An equation relating the height of the water-table between drains to depth of the drains and the distance between them, the hydraulic conductivity of the soil, and the rate of recharge by percolating water is called:

(*a*) Hooghoudt equation (*b*) Darcy equation

(*c*) Renould's equation (*d*) None of the above

Q. 62 The volume of water that runs off the soil surface and flows into streams during episodes of intense rainfall is called:

(*a*) Runoff (*b*) Erosion
(*c*) Splash erosion (*d*) Sheet erosion

Q. 63 The volume of water that runs off during an irrigation that exceeds the soil's infiltrability is called:

(*a*) Run-off (*b*) Erosion
(*c*) Splash erosion (*d*) Sheet erosion

Q. 64 LCC (Land capability class) 1 is denoted by which colour?

(*a*) Light green (*b*) Yellow
(*c*) Red (*d*) Blue

Q. 65 LCC (Land capability class) 2 is denoted by which colour?

(*a*) Light green (*b*) Yellow
(*c*) Red (*d*) Blue

Q. 66 The erodability increase at what size of aggregates?

(*a*) 0.2mm (*b*) 0.3mm
(*c*) 0.4mm (*d*) 0.5mm

Q. 67 Which is the most important index of soil erodibility?

(*a*) Silica content (*b*) Sesquioxide content
(*c*) Silica-sesquioxide ratio (*d*) All the above

Q. 68 Each orbital plane of satellite in FPS is inclined at - equatorial plane of the earth:

(*a*) 55 degree (*b*) 45 degree
(*c*) 35 degree (*d*) 60 degree

Q. 69 The orbital plane of the satellite is displaced by which of the following longitude:

(*a*) 100 degree (*b*) 110 degree
(*c*) 120 degree (*d*) 90 degree

Q. 70 NAVSTAR was launched in the year:

(*a*) 1978 (*b*) 1968
(*c*) 1958 (*d*) 1982

Q. 71 Which of the following GPS satellite is planned to be launched by India:

(*a*) NAVSTAR (*b*) GLONAS

(*c*) GAGAN (*d*) None

Q. 72 P codes and C/A codes are:

(*a*) Pseudo random binary codes

(*b*) Random binary codes

(*c*) Pseudo random logarithmic code

(*d*) Random logarithmic code

Q. 73 Which of the following represents features that have a known location on earth:

(*a*) Spatial Data (*b*) Attribute Data

(*c*) Data Layers (*d*) Layer Type

(*e*) Topology

Q. 74 The information linked to the geographic features (spatial data) that describe those features is:

(*a*) Spatial Data (*b*) Attribute Data

(*c*) Data Layers (*d*) Layer Type

(*e*) Topology

Q. 75 How geographic features are related to one another and where they are in relation to one another is known by:

(*a*) Spatial Data (*b*) Attribute Data

(*c*) Data Layers (*d*) Layer Type

(*e*) Topology

Q. 76 Spatial or coordinate data represents features that have a known location on the earth in X & Y locations:

(*a*) Points (*b*) Polygon

(*c*) Both (*d*) None

Q. 77 Data input in GIS can be made in the form of:

(*a*) Vector data (*b*) Raster data

(*c*) Alpha numeric data (*d*) All

Q. 78 IRS- 1C/1D PAN spatial resolution is:

(*a*) 5.8 m (*b*) 24 m

(*c*) 188 m (*d*) 30 m

Q. 79 PAN camera of IRS 1C/1D provide which of the following type of image:

(*a*) Coloured (*b*) Blue and red

(*c*) Blue and green (*d*) Black and white

Q. 80 Repeatitivity of IRS- 1C/1D satellite is:

(*a*) 16 days (*b*) 24 days

(*c*) 3 days (*d*) 1 hour

Q. 81 In which of the following condition reflectance % will be maximum:

(*a*) Dry soil (*b*) Wet soil

(*c*) ploughed soil (*d*) Water bodies

Q. 82 What could be the reflectance % if the farmer is applying FYM in their field:

(*a*) Increased

(*b*) Decreased

(*c*) No effect

First increases then decreases

Q. 83 Which of the following kind of wave is related to atmospheric window:

(*a*) Long wave (*b*) Short wave

(*c*) Both (*d*) None

Q. 84 Which of the following data structure can best describe coordinate transformation:

(*a*) Vector data structure (*b*) Raster data structure

(*c*) Both (*d*) None

Q. 85. Stratosphere extends up to a height of:

(*a*) 8–18 km (*b*) 18–50 km

(*c*) 50–80 km (*d*) >80 km

Q. 86 Conduction is:

(*a*) The transfer of heat by molecular activity

(*b*) The process of heat transfer within liquids and gases resulting from the motion of fluids

(*c*) The transfer of energy by electromagnetic waves

(*d*) all

Q. 87 The quantity of water vapour present in the atmosphere at any particular time is expressed as:

(*a*) Absolute humidity (*b*) Specific humidity

(*c*) Relative humidity (*d*) All

Q. 88 Absolute humidity is:

(*a*) Weight of water vapour/ volume of air

(*b*) Weight of water vapour/ weight of air including water vapour

(*c*) Water vapour present in the atmosphere/ water vapour required for saturation × 100

(*d*) None

Q. 89 The instrument used for measuring relative humidity is:

(*a*) Psychrometer (*b*) Hygrometer

(*c*) Both (*d*) None

Q. 90 The wet and dry bulb thermometers are kept in:

(*a*) Side screen (*b*) Sight screen

(*c*) Stevenson's screen (*d*) Sun screen

Q. 91 RH can be accurately measured in crop canopy with:

(*a*) Assman's psychrometer (*b*) Hair hygrometer

(*c*) Hygrograph (*d*) None

Q. 92 The diameter of cylindrical vessel in sunken screen evaporimeter is:

(*a*) 4.5 cm (*b*) 45cm

(*c*) 60 cm (*d*) 60 mm

Q. 93 The electronic energy levels of the hydrogen atom in the Bohr theory are called as:

(*a*) Orbitals (*b*) Orbits

(*c*) Rydberg levels (*d*) Ground states

Q. 94 A node is a surface on which the probability of finding an electron is:

(*a*) 0 (*b*) >1

(*c*) >10 (*d*) >90

Q. 95 Bohr's model of an atom is based on:

(*a*) Quantum theory

(*b*) Dalton atomic theory

(*c*) Theory of electrolytic dissociation

(*d*) Law of mass action

Q. 96 Artificial radioactive was discovered by:

(*a*) Madame Curie

(*b*) Fermi

(*c*) Irene Curie and Federic Joliot

(*d*) Rutherford

Q. 97 Dipole moment can be measured by determining ....:

(*a*) Resistance (*b*) Potential difference

(*c*) Dielectric constant (*d*) Current

Q. 98 Dielectric constant can be measured from the ratio of the electrical capacity of ......:

(*a*) Condenser filled with substance to empty condenser

(*b*) Empty condenser to filled condenser with substance

(*c*) Both

(*d*) None

Q. 99 Which of the following solvent(s) is/are non-polar:

(*a*) Benzene (*b*) CCl_4

(*c*) Both (*d*) None

Q. 100 H-bonding, taking place between the side chains of amino acids, is one of the forces responsible for the folding of which of the following into tertiary or quaternary structure of protein:

(*a*) α-helix (*b*) β-helix

(*c*) γ-helix (*d*) None

1	(*b*)	23	(*b*)	45	(*d*)	67	(*d*)
2	(*b*)	24	(*b*)	46	(*d*)	68	(*a*)
3	(*c*)	25	(*d*)	47	(*d*)	69	(*c*)
4	(*d*)	26	(*b*)	48	(*d*)	70	(*a*)
5	(*a*)	27	(*c*)	49	(*d*)	71	(*c*)
6	(*c*)	28	(*d*)	50	(*e*)	72	(*a*)
7	(*d*)	29	(*c*)	51	(*b*)	73	(*a*)
8	(*a*)	30	(*a*)	52	(*d*)	74	(*b*)
9	(*c*)	31	(*a*)	53	(*a*)	75	(*e*)
10	(*a*)	32	(*d*)	54	(*a*)	76	(*a*)
11	(*b*)	33	(*b*)	55	(*c*)	77	(*d*)
12	(*d*)	34	(*e*)	56	(*d*)	78	(*a*)
13	(*c*)	35	(*c*)	57	(*d*)	79	(*d*)
14	(*c*)	36	(*a*)	58	(*d*)	80	(*b*)
15	(*b*)	37	(*a*)	59	(*a*)	81	(*a*)
16	(*a*)	38	(*d*)	60	(*c*)	82	(*b*)
17	(*b*)	39	(*b*)	61	(*a*)	83	(*a*)
18	(*a*)	40	(*c*)	62	(*a*)	84	(*b*)
19	(*d*)	41	(*b*)	63	(*a*)	85	(*b*)
20	(*c*)	42	(*d*)	64	(*a*)	86	(*a*)
21	(*b*)	43	(*c*)	65	(*b*)	87	(*d*)
22	(*a*)	44	(*b*)	66	(*a*)	88	(*a*)

89	(*c*)	92	(*b*)	95	(*d*)	98	(*c*)
90	(*c*)	93	(*b*)	96	(*c*)	99	(*c*)
91	(*b*)	94	(*a*)	97	(*c*)	100	(*a*)